BEYOND MIND–

BEYOND MIND–
THE MILELA THEORY

❖ ❖ ❖

CHETANA YOGA

The Next Step in Human Evolution

Dr. Rajah Sekaran

DR. RAJAH SEKARAN

First edition published in 2019
Second edition revised and puplished in 2020
M. Rajah Sekaran, Member, Yogesha, LLC
Email: mrajahsekaran@gmail.com

First edition published in 2019
Second edition revised and puplished in 2020
M. Rajah Sekaran, Member, Yogesha, LLC
Email: mrajahsekaran@gmail.com

First edition published in 2019
Second edition revised and puplished in 2020
M. Rajah Sekaran, Member, Yogesha, LLC
Email: mrajahsekaran@gmail.com

Cover design by Elite Editing

© M. Rajah Sekaran 2020
All rights reserved. This book is protected under the copyright laws of the United States of America and other countries. Except for reviews and quotations, use or republication of any part of this work is prohibited, without prior written permission from the author who holds the copyright.

ISBN-13: 978-0-692-19861-2

ISBN-13: 978-1975602475

Library of Congress Copyright Office in accordance with Title 17, *United States Code* (under the title *Origin of All Things in Nature*—year 2015)

Registration Number: TXU1-998-639

PLEASE NOTE

M. Rajah Sekaran, Member, Yogesha LLC, the author and publisher disclaim any implied warranty, guarantee, liability, loss or injury in connection with any of the information or suggestion, described in this book.

The author and publisher are not engaged in rendering medical or any other kind of personal, professional services in this book. You are advised to consult appropriate medical or professional help for all matters related to health, before using the information and suggestions given in this book.

CONTENTS

Prologue xix

Section One:
The Eternal Dance 3
        A Collection Of Poems 5

Section Two:
Chetana Yoga 17
  Chapter 1 Beyond Mind 21
  Chapter 2 Chetana Yoga— The Next Step
                  In Human Evolution 33

Section Three:
Epiphany 47
  Chapter 1 Epiphany 49
  Chapter 2 Limitations And Limitlessness 65

Section Four:
Milela Theory **71**
    Chapter 1   The Milela Theory 73
    Chapter 2   Life Energy of Milela 85
    Chapter 3   Limitless Awareness 107

Section Five:
Origin **127**
    Chapter 1   Space And Subtlety 129
    Chapter 2   Origin of all Things in Nature ... 149

Section Six:
Mind Management **155**
    Chapter 1   The Body-Mind Connection 157
    Chapter 2   The Brain-Mind Connection 165
    Chapter 3   Mind Management 179

Section Seven:
Meditation **195**
    Chapter 1   Pranayama 197
    Chapter 2   On Mediation 213

Epilogue 223

Chetana Yoga, by being Awareness-based, creates a permanent layer between the mind and the action; and gives an opportunity to evaluate the action instantaneously, before it happens. Thus, it strengthens the Free-Will to choose the right action—either right communication or right action with good intentions—and keeps the mind as a life-long friend.

The transliteration and/or translation of Sanskrit, Tamil and Latin words, used in this book are mostly given following the words, in the text itself.

Glossary is given at the end of the book.

Dedicated to All my Teachers and to my family from whom I learned many lessons on life

To see a World in a Grain of Sand
And a Heaven in a Wild Flower,
Hold Infinity in the palm of your hand
And Eternity in an hour

If the doors of perception
Were cleansed
Everything would appear to man
As it is, infinite.

–William Blake
(1757-1827)

DR. RAJAH SEKARAN is a California-based retired physician who completed his medical training in both his native India and in the United States. He has been an avid student of philosophy and the sciences, studying both Eastern and Western religions for nearly 40 years. During his studies, Dr. Sekaran realized that while humanity has made advances in technology and social comforts, all we have done in terms of violence is replace bows and arrows with guns and bombs. Part of the inspiration for this book was the realization that with new discoveries in the field of quantum physics, ancient spiritual beliefs may hold answers to some of modern science's deepest unanswered questions.

Dr. Sekaran has three grown children and lives in Berkeley, California, with his wife.

PROLOGUE

THE REVELATIONS, DESCRIBED IN THIS book, occurred to the author recently as epiphany. However, these revelations are a culmination of nearly 40 years of meditation, deep contemplation, along with the study in Sanskrit, Tamil, and English of different philosophies of spirituality and religions, and also reinforced by a basic knowledge of Modern Physics, also known as Quantum Physics. The contents of this book are not mere reproduction of knowledge from other studies or books. They are mostly *realizations* from experience and that is the reason I am calling them *revelations*.

In the olden days, in India, renunciates known as Sannyasis were a recognized division of the society. The Sannyasi order was supported by the society, in order to explore the secrets of *Reality, Truth and Nature;* and to teach the same to others. However, in the modern world, I had to be a house-holder to meet my family's obligations and continued the pursuit of *the Ultimate Truth*, in spite of many difficulties and obligations.

A house-holder's life, however, is a life of *possessions* by necessity and becomes mind-based. Maturity comes when we can live a life of possessions and still be *beyond-mind in an Awareness-based life*. That is the subject of this book. On the other hand, even a renunciate needs at the least, a few *possessions* to survive, such as a supply of food and clothes.

It is ironic that the trials and tribulations of family life are what make a person mature by giving opportunities to learn the lessons of life and to purify the mind. Purified mind and deep meditation and contemplation are the pre-requisites for the realization of *Ultimate Absolute Truth and Reality*.

However, the practice of *Chetana Yoga* does not necessarily require the pre-requisites of deep meditation and realization of *Truth and Reality*, as will be explained in the chapter on Chetana Yoga.

Awareness in *Sanskrit* is variously described as *chith, chita, chit-akasha, chetana and chaitanya*. In both Sanskrit and *Tamil*, another classical language of India, it is combined with *sat* (truth or existence) and *ananda* (bliss) and is given the name *sat-chit-ananda*. Gautama Buddha 2,500 years ago called the Awareness in his mother tongue, *Pali*, as *Chetana*.

Awareness is a synonym of *Consciousness*. In the Western World, as well as in the East, the word

Consciousness is used to indicate Awareness, and extensive research is being done by both philosophers and physicists all over the world to analyze, understand, define, and prove the nature of *Consciousness*.

Philosophers and yogis are seeking to understand and realize *Consciousness*.

Whereas, the physicists are trying to prove its existence mathematically from Physics' point of view. However, the word *Consciousness* often leads one also to the concept of *unconsciousness* and *subconscious* etc., which are confined to the mind creating confusion. Some philosophers of the past have dealt with this confusion created by the mind, as mentioned above, by adding another dimension, known as the state of *super-consciousness*.

Another philosopher, Saint Ramalinga Swami, who lived in India in the nineteenth century, has described *supra-mental* approach as the method to realize Awareness. Saul-Paul Sirag from the Institute for the study of Consciousness in Berkeley and Parapsychology Research group in San Francisco "has developed a hyperspace model of Consciousness, describing mathematical structures called reflection spaces, which are hierarchically organized in such a way that *infinite* spectrum of realities is naturally suggested".

Similarly, physicists have described the space-time aspect of *Consciousness* as hyper-dimensional, with all but four of the dimensions being invisible.

To avoid the ambiguities created by the mischievous mind, I have chosen to use Awareness as the preferred word, instead of *Consciousness*, even though they are synonyms. Furthermore, I have given proof of the presence of Awareness, *even in the absence of mind*, as in *deep sleep and fainting spells*, (described in detail in the subsequent chapter under the title "The *MILELA Theory.*")

At this stage of human evolution, I find, it is an absolute necessity to evolve from a *mind-based life* to *an Awareness-based life*. This immediate need for the next step in human evolution is further enhanced by the discouraging reports from different news media, of *increasing gun violence, suicide and suicide attempts* and the rise of *loneliness and depression, especially in the teenage children* in America as well as in other parts of the world.

Gun Violence, Suicide and Suicide Attempts and Loneliness and Depression are Only Some of the Visible Symptoms of a Basic Mental Malfunction
The Tip of a Titanic Type of Iceberg

The rising rates of gun violence reports from news media and the suicide reports and non-fatal self-harm reports from CDC (Centers for Disease Control and prevention) quoted below, and loneliness and depression among the general population, all appear to be the *tip of a titanic type of iceberg*, which very well may represent millions of others in the general population with much milder forms of different degrees of mental malfunction.

This mental dysfunction is sub-clinical and may not qualify as a mental disease.

Gun violence and suicide and non-fatal self-harm and depression are only some of the severe symptoms of a *basic malfunction of mind* showing up as *increasing aggression and prejudice and/or loneliness* in the minds of men and women and children all over the world. The same malfunction (not considered as a mental disease) is seen in milder form in millions, in everyday life in USA and also in the rest of the world. Domestic violence, sexual harassment and milder 'type1' bipolar disorder and loneliness or depression are not uncommon in the general population in USA, as well as in other parts of the world. and often go unnoticed.

RECENT EVIDENCE OF INCREASING GUN VIOLENCE–MASS SHOOTING

I have not included all of the gun violence and other violence using knives etc. Only the mass shootings are included in this meta-analysis, mainly to show *the trend in the increase of violence* and aggression in the minds of men, women and children.

Guardian U.S. Interactive Team reports as of February 15th, 2018 from January 1st, 2013, in the past five years, there have been 1,875 deaths and 6,848 injured in 1,870 days, from mass shootings alone. In the 50 years before the *Texas Tower Shooting* on August 1st, 1966, there were just 25 public *mass shootings* in which four or more people were killed according to a criminologist, Grant Duwe, author of "Mass Murder in the United States: A History." Since then, his report compares it to the next 50 years from 1966 to 2018, and the number has risen dramatically to 154 mass-shootings with a lot more casualties; and many of the deadliest shootings have occurred within the past few years.

The number of injured has been much greater than the number of deaths. The loved ones affected by these incidents are too many to count.

Dave Mosher and Skye Gould reported in Business Insider that gun violence is a leading cause of death in America, based on 2015 National Safety Council, National Center for Health Statistics. The odds of death from varying causes were compared; and from assault by gun violence was 1 in 315 and from mass shooting (defined as four or more victims) was 1 in 11,125 and from foreign born terrorists (based on 41-year average from 1975 to 2015) was 1 in 45,785.

In TIME Magazine, Chris Wilson updated as of November 5, 2017 and quoted a database compiled by

'Mother Jones', going back to 1982 and included only mass shootings in which at least three people were killed, not including the gunman. In that time, 722 people have been killed and 1,177 wounded. There were 91 such incidents.

On my review of all such incidents, from the report above, I found that from 1985 to 2004 there was mostly *one* shooting per year with the number of people killed in the *single* digit. After 2005, there were mostly *two* incidents of mass shooting per year and the number of people killed was in the *double* digits or more. In 2012, the number of mass shootings increased to *seven* and 67 people were killed and 68 were wounded. The number of mass shootings have become more and more frequent. In 2017 there were *ten* mass shootings with 112 killed and 531 wounded; and the frequency continues to increase.

The *New York Times* report by Michael S. Schmidt on September 24th, 2014 says "F.B.I. confirms a sharp rise in mass shootings since 2000"–"there were, on average, 16.4 such shootings a year from 2007 to 2013, compared with an average of 6.4 shootings annually from 2000 to 2006. In the past 13 years, 486 people have been killed in such shootings, with 366 of the deaths in the past seven years. In all, the study looked at 160 shootings since 2000. (Shootings tied to domestic violence and gangs were not included.)"

Are the Mass-Shooters Mentally Ill?

The *Washington Post* report by Bonnie Berkowitz and others, updated as of June 29th, 2018 analyzed the details of each shooter and found only some of the mass shooters were known to have violent tendencies or criminal pasts. Others seemed 'largely fine' until they attacked. All but three were male. The vast majority were between the ages of 20 and 49.

In summary, my review of several reports of all shootings, including mass shootings, revealed increasing incidence of gun violence in the 21st century compared to the last quarter of the 20th century. Moreover, a review of the shooters reveals only some of them had history of mental illness, violent tendencies, or criminal pasts. The majority of the shooters were men between the ages 20-49, and they were "largely fine" until they attacked. The general population has become *numb* to the news of gun violence, drug-related shootings and suicides; and is beginning to consider them as something similar to automobile accidents. The news media also now-a-days is pushing such news to the back pages; they used to be front-page news, often as headlines. I am not finding fault with their journalism; but that is the way it is.

Gun Violence in USA Compared to other Countries

A report from NPR (National Public Radio) compared the other developed countries where violent gun deaths per 100,000 people in 2016 ranged from 0.03 in Singapore to 0.48 in Canada. In the United Kingdom it was 0.07. In comparison, in the United States, it was 3.85— about *30x* more than Germany which had 0.12, and *27x* higher than Denmark which had 0.14, and *8x* higher than the rate in Canada. Among the countries in the East, in India for example, there are 0.88 deaths per 100,000 people, which is about one fourth of the gun deaths per capita in the United States even though India has the second largest civilian firearm stock pile, estimated at 46 million with a population that is about four times that of USA. According to a report from the National Institute of Justice (NIJ) in 2009, the number of firearms among American civilians is estimated around 310 million, about average one per person.

SUICIDE RATES RISING: CDC PRESS RELEASE

CDC (Centers for Disease Control and Prevention) in its 'Vital Signs Report' as the 'Press Release' of June 7th 2018 says, "suicide rates have been rising in nearly every state. Twenty-five states had suicide rate increases of more than 30 percent."

CDC researchers—Deborah M. Stone, ScD; Thomas R. Simon, PhD; Katherine A. Fowler, PhD; Scott R. Kegler, PhD; Keming Yuan, MS; Kristin M. Holland, PhD; Asha Z. Ivey-Stephenson, PhD; Alex E. Crosby, MD— further report on June 8th 2018 and parts of their detailed report is given here, in their own words, "Suicide rates in the United States have risen nearly 30 percent since 1999, during 1999-2016. 54 percent of decedents in 27 states in 2015 did not have a known mental health condition. In 2016, nearly 45,000 Americans age 10 or older died by suicide. Suicide is the 10th leading cause of death and is one of just three leading causes that are on the rise.

In addition, rates of Emergency department visits for non-fatal self-harm, increased 42 percent from 2001–2016. Together, suicides and self-harm injuries cost the nation approximately $70 billion per year in direct medical and work loss costs (the cost quoted here mainly to show the magnitude of the problem).

Firearms were the most common method of suicide overall (48.5%); decedents without known mental health conditions were more likely to die by firearm (55.3%).

Suicide is rarely caused by a single factor, but rather, is determined by multiple factors. In addition to mental health conditions and prior suicide attempts, other contributing circumstances include social and economic problems, access to lethal means (e.g., substances, firearms) among persons at risk, and poor coping and problem-solving skills."

LONELINESS

The Economist magazine, in its September 2018 issue, published an article on loneliness, under the title, "Alone in the crowd—loneliness is increasingly being treated as a serious public-health problem".

Loneliness is defined as the subjective feeling of isolation from family and friends. Loneliness is different from solitude, when a person chooses to be alone. Using a questionnaire from UCLA (University of California, Los Angeles), a study published in 2010, reported that about 35% of Americans over 45 were lonely. Vivek Murthy, former Surgeon General of USA in 2017, compared loneliness to obesity and smoking, as an *epidemic*.

Julianne Holt-Lunstad of Brigham Young University, in Utah did a meta-analysis in 2015 analyzing 70 papers with 3.4 million participants over an average of seven-year period and found that those classified as 'lonely' had 26% higher risk of dying and those living alone had 32% higher risk, adjusted for age and health status.

ONS (Office for National Statistics) in UK in 2013, using a single question, found 25% of people, over 52 years old, were 'sometimes lonely' and another 9% 'often lonely'. British government this year created a cabinet post for 'loneliness.'

The Economist article further added, "Other smaller scale studies show loneliness leading to other health problems, such as heart attacks, strokes, cancers, eating disorders, drug abuse, alcoholism, depression and anxiety."

Depression

Johns Hopkins Health Review, in 2017 published an article on the rising incidence of teen-age depression and reported, "According to a study by Ramin Mojtabai, professor at Johns Hopkins Bloomberg school of Public Health, adolescents suffering from clinical depression grew by 37% between 2005 and 2014. The National Institute of Mental Health estimates that three million adolescents, ages 12 to 17 have had at least one major depressive episode in the past year (2016-2017). Teen depression appears to be on the rise equally among urban, rural and suburban populations. Research also shows that more dangerous behaviors, like self-harm, are increasing."

Malfunction of Mind on a Global Scale

Such pathological aggression and/or loneliness and depression, in the mind shows up in different forms in the different parts of the world. It only appears as gun violence in USA because of the availability of guns and easy access to guns from the gun stores as well as at home with irresponsible storage of guns, in USA.

On a global scale, it also appears as *oppression* of another race, religion, caste or sub-caste or tribe, leading to violence causing some whole ethnic communities having to move out of their home-lands. Or one country may decide to aggressively occupy another country.

It appears as *oppression of women* in some countries. Or it may appear as suicide-bombing, we have witnessed in some parts of the world due to ideological differences. Religious wars of the past are also examples of such pathological aggression and oppression in the minds of men and women.

That is why we are reaching a stage of acute need for a *paradigm shift* from a mind- based life to *Awareness-based life* that is within our reach, through the *Chetana Yoga* described in this book, reinforced by *proof* that *Chetana* meaning the Awareness is *beyond mind* and exists even when the mind is shut off as in deep sleep or in a fainting episode.

The Paradigm Shift

This is my humble attempt to introduce a *Paradigm shift* as a remedy for the mental frustrations and loneliness and for the increasing violence in the minds of men, women and even children of high-school age; and in that attempt, at the same time, to reconcile Philosophy and Physics, Religion and Science, Creation and Evolution, Mysticism and Quantum Physics, *by proposing and reconciling* in this book, three related *paradigm-shifts*, as follows:

1. *CHETANA YOGA*
2. THE *MILELA THEORY* (Milela stands for Mutually Inherent Life Energy and Limitless Awareness) and
3. SPACE AND *THE THEORY OF SUBTLETY*

1. CHETANA YOGA–Paradigm Shift

Such a paradigm-shift in the *individual level*, is an *evolution* from *Body-Mind orientation of life to Life Energy–Awareness orientation* which is *Limitless* and more *fundamental* than the body-mind existence which is *limited* by nature. It is the *limited mind* that suffers from the problems of anger, anxiety, vanity, arrogance and violence in the *individual level* and portrays the racial, regional and caste prejudices as a *society*.

Along with the Paradigm shift from body-mind existence to the *Life Energy-Awareness existence*, I am also describing the method of living such a transformed existence. The method is named as *Chetana Yoga*.

It is important to know that the mind does not disappear in this paradigm-shift from mind-based existence to Awareness-based existence. The mind gains *clarity and focus*, now devoid of the frustrations due to anger, anxiety, stress and/or prejudices. For example, the hand does not disappear, when you are being aware of the hand and its actions.

To correct the problems of the modern world such as generally increasing mental frustrations due to anger, aggression and anxiety and loneliness in the *individual level*, and the use of guns and bombs and drug-related crimes at the *society level*, and the problems in the *large-scale level* due to racial, religious and regional prejudices, *Chetana Yoga-living* is needed by necessity, perhaps with a teleological purpose for the humanity in general, as *the natural progression of human evolution from mind-based life to Awareness-based Life.*

2. THE MILELA THEORY–Paradigm Shift

MILELA Theory is at the center of the other two paradigm shifts, being the very basis of the *Chetana Yoga and Theory of Subtlety*. I have explained a few *thought experiments as observational inferences* to prove the *theory of Life-Energy being mutually inherent in Awareness*, e.g., the absence of mind during deep sleep and fainting spells, in one of the descriptions.

Milela Theory is also presented as a *General Theory of all things in Nature.*

3. SPACE and THE THEORY OF SUBTLETY–Paradigm Shift

The complexity of the inspirations in formulating the *paradigm shifts* was also compounded by my desire as well as the need and teleological purpose in the human evolution, to reconcile the world of Quantum Physics with our macroscopic world. In the *quantum physics level* also, this paradigm-shift to *Life Energy–Limitless Awareness-Space* is more *fundamental* than the existence of gravity, electro-magnetism, sub-atomic particles as well as the billions of galaxies, blackholes and quasars and this world.

Addendum to the Prologue
Remember the Titanic

Herein the author is further expanding the discussion on the generalized mental malfunction leading on to gun violence, suicide, suicide attempts, loneliness, and depression. These incidents were described in the Prologue as the tip of the Titanic-like iceberg, wherein the subclinical malfunction of the mind is the deep-seated iceberg.

Incidence of ADHD (Attention Deficit Hyperactivity Disorder) and Pollution

Since 1997, the incidence of ADHD has nearly doubled, according to The Center for Disease Control and Prevention. Annual Reports from the National Health Interview Survey report an increase from around 5% in 1997 to nearly 10% in 2017. Multiple factors seem to be involved in the etiology of ADHD. Both genetic and environmental factors are associated with the increased incidence of ADHD. Among the environmental factors, smoking during pregnancy and exposure to lead in children have been proven to cause the increased incidence. However, the effect of pollution, both air pollution and other forms of pollution, are still not studied. More research is needed to draw a correlation between pollution and the incidence of ADHD in children.

Air Pollution and Neuro-Psychiatric Disorders

There is a growing field of study researching the effects of air pollution on neuro-psychiatric disorders. In August 2019, University of Chicago study published a study about the association between air pollution and neuropsychiatric disorders including bipolar, major depression, schizophrenia, and personality disorder in PLoS Biology. The study focused on two populations, the US and Denmark, and four disorders: bipolar, major depression, schizophrenia, and personality disorder. The study included data from 151 million people from the US, and 1.4 million people from Denmark. It was interesting to note that in spite of the differences in the demographics, healthcare practices and social structures the results were very similar. Both studies found a higher incidence of bipolar disorder in populations with high air pollution. In the US, the counties with the most air pollution had a 27 percent increase in bipolar disorder when compared to the population with the least air pollution. The results from Denmark were almost the same, with a 29 percent increase in the incidence of bipolar disorder associated with the worst air pollution. Additionally, the study in Denmark showed increased incidence of schizophrenia and personality disorder, whereas the US study mainly focused on bipolar disorder and major depression.

We are now encountering more and more evidence of the relationship between air pollution and changes in the brain. There is a metanalysis study by Paula de Prado Bert et al. from Pompeu Fabra University of Barcelona, Catalonia, Spain and MRI Research Unit, Department of Radiology, Hospital del Mar Barcelona, Spain etc., titled "The Effects of Air Pollution on the Brain: a Review of Studies Interfacing Environmental Epidemiology and Neuroimaging."

The metanalysis focused on 11 studies, on adults and children, in both urban and rural communities. The studies included children from Mexico City and included Neuroimaging investigation of 40 children from New York City, revealing a dose-response relationship between increased prenatal hydrocarbons exposure and significant reduction of the cerebral white matter. "The effect of TRAP (Traffic Related Air Pollution) on cognition appears to be biologically plausible.... Moreover, findings from all these studies suggest that, because of their localization and their magnitude, these air pollution-related brain changes visualized in vivo could mediate the effects of air pollution on cognition."

While the studies found no strong correlation between exposure to pollution and effects to subcortical brain regions (grey matter), the studies did reveal a correlation between pollution and volumes of white matter in the brain. White matter isn't as critical to our thoughts and actions as grey matter, which compromises regions such as the amygdala and hippocampus, which are critical for everyday functions such as memory and emotion-processing. However, white matter is still essential for brain functioning, and the volume of white matter is especially important in the elastic brains of children. One study revealed that children in Mexico City experienced differences in the right parietal and bilateral temporal areas, when compared to controls. When comparing the children in Mexico to the control group, the experimenters found deficits in attention, short-term memory, and learning abilities associated with the urban environment. Moreover, neuroimaging of children from New York City revealed a correlation between prenatal PAHs (Polycyclic aromatic hydrocarbons) exposure and significant reduction in the cerebral white matter covering nearly the entire left hemisphere of the brain. Additionally, long term exposure to air pollution has been shown to adversely affect the general population. The negative impacts are observable using various MRI modalities.

Pesticide Pollution: Glyphosate and Chlorpyrifos

It is well-known and has attracted media attention with hundreds of millions in punitive damages paid to the plaintiffs by the manufacturer of glyphosate (Round-Up Herbicide). The large sums of money were awarded to the users of glyphosate, because they developed cancer— non-Hodgkin's lymphoma— and the jury found a connection between the cancer and the herbicide.

In May 2004, 140 agricultural workers were told by their supervisors to flee the field after pesticide fumes from an adjacent field sickened 19 of them. So, California proceeded to ban the pesticide Chlorpyrifos effective February 6, 2020, and the farmers will still be able to exhaust their supplies ntil the end of 2020. California environmental officials declared the pesticide Chlorpyrifos as being linked to brain damage in children.

Micro-plastics and Pollution

Micro-plastics found in the ocean have increased multifold, polluting the marine life and harming them. Now, there is also evidence, the micro-plastic particles seen in bottled water, tap water, beauty products, and food containers are harming the humans as well, besides the ingestion of seafood and other marine life by the humans.

It is commonly known that micro-plastic particles are harming marine organisms, and Chelsea Rochman, a professor of ecology at the University of Toronto, conducted a study to test the effects of soaked ground-up polyethylene, a product used to make some types of plastic bags on a small fish species known as Japanese medakas. She found that the fish who ingested micro-plastic particles from the polyethylene were more likely to have liver damage. Moreover, another experiment found that polystyrene— another common micro-plastic, derived from take-out food containers— caused the fish to produce fewer eggs and sperm with low motility.

In various scientific experiments, micro-plastics have been found in over 114 marine species, and over 50% are commonly eaten in our homes. Still, more research is needed to determine the effects on human health.

An article in *The National Geographic* explains the importance of chemical impacts of micro-plastics on our everyday lives. "In addition to mechanical effects, microplastics have chemical impacts, because free-floating pollutants that wash off the land and into our seas—such as polychlorinated biphenyls (PCBs), polycyclic aromatic hydrocarbons (PAHs), and heavy metals—tend to adhere to their surfaces... It's difficult to parse whether microplastics affect us as individual consumers of seafood, because we're steeped in this material—from the air we breathe to both the tap and bottled water we drink, the food we eat, and the clothing we wear. Moreover, plastic isn't one thing. It comes in many forms and contains a wide range of additives—pigments, ultraviolet stabilizers, water repellents, flame retardants, stiffeners such as bisphenol A (BPA), and softeners called phthalates—that can leach into their surroundings."

SOLUTIONS TO THE ICEBERG PROBLEM
In the discussion on gun violence, suicide, suicide attempts, loneliness and depression, they were previously

grouped together as the tip of the iceberg. The iceberg itself was described as the widespread mental malfunction, because all actions including the above incidents described as the tip of the iceberg, arise from the minds of people. So, such actions depend on the status of the mind.

The above studies regarding different types of pollution correlate the environmental factors and pollution to the mental status as well as the general health of the persons involved. So, what is the solution?

What is the Solution?

Considering the issues discussed above, the solution needs to be directed from three different approaches. The three-pronged treatment, by necessity, needs to involve:

1. Gun-Control or Gun-Management
2. Pollution-Control or Environmental-Management
3. Mind Management.

1. GUN-CONTROL OR GUN-MANAGEMENT

a. By necessity, needs to involve legislative changes such as age limits, background checks, delayed delivery of the weapon, and removal of military style assault weapons such as AK-47 from the general circulation

b. Education of gun-owners: The gun-owners need to be educated about safekeeping and avoiding access to the children.
c. There is also an argument that even if the legal possession of firearms is well controlled, the illegal acquiring and possession of firearms is beyond the control of legal system. So, the intelligence departments need to be aware of the issue and extensive measures need to be availed

2. POLLUTION-CONTROL OR ENVIRONMENTAL-MANAGEMENT
   The Paris Agreement and the Conference on Climate Change have identified USA and China as the Top 2 polluters of the world, with a combined total of 45% of total worldwide emissions. The World Economic Forum estimates the pollution by the top 5 polluters as follows:

   China: 30%
   USA: 15%
   EU: 9%
   India: 7%
   Russia: 5%.

Nevertheless, as the pollution problem is global, the solutions also have to be from the global participation of all countries. Studies have shown that we are very close to the irreversible stage of climate change.

So, we have now reached a stage where solutions have to be implemented immediately before the climate change becomes irreversible.

The United Nations: In the seventy-third session of a high-level meeting on climate change and sustainable development, on March 28, 2019, The United Nations warns the global population that we are the last generation that can save the globe from irreversible climate change. Further, the speakers warned there are only 11 years left before we reach the point of irreversible climate change. An article in *The National Geographic* claims that "there is a real possibility that we will be entering a phase of accelerated Arctic warming in the next two to four decades if mitigation action isn't taken soon," says Post, a climate change ecologist at the University of California, Davis…The Arctic is warming far more quickly than anywhere else on the planet. Temperatures climbed nearly 1.8 degrees Fahrenheit (1 Celsius) in the past decade alone. At the current rate of greenhouse gas emissions, the North is on track to warm 7.2°F (4°C) year-round—and top 12.6°F (7°C) in autumns—by the middle of this century, according to the report. That's about when the planet as a whole is projected to reach the 3.6°F (2°C) warming often cited as the threshold for disastrous impacts."

Prologue

An article in *Scientific American* from May 2017 describes studies conducted in Antarctica, that conclude warming trends are similar to changes in the Arctic region. Researcher Matt Amesbury from the University of Exeter in the United Kingdom, and his fellow researchers, used samples of a moss bank to reach their conclusion: the warming in the Antarctic regions is allowing the growth of plant life, such as phytoplankton.

With the background of studies given above, the solutions need to include immediate action in reducing the global carbon pollution. It includes complete replacement of gas-powered engines for all cars and trucks to hybrid or electric-powered engines globally. Reductions in the use of pesticides, plastics and microplastic particles in the everyday life need to be brought about with legislative measures in all the countries.

IF THERE IS A WILL, THERE IS A WAY.
There is no need to panic. There are on-going developments and innovations that can help the pollution problem. However, willingness is needed on the part of the governments, politicians, and technology firms from the private sector. For example, to fight carbon pollution, planting trees is an obvious solution, but it is a long-term process. Besides, other recent innovations for the immediate future include:

1. Algae to reduce carbon pollution
2. Mealworms that can consume plastics and Styrofoam.

Algae to reduce carbon pollution:
Ben Lamm, Founder of a firm developing the use of algae, writes in October 2019 in QUARTZ Online Publication, "Algae can be utilized in a number of ways to reduce carbon in the atmosphere. Other than it being the most efficient solution for storing carbon dioxide, it can be easily used in a variety of other sutainable and commercial products or materials, from tennis shoes to steel alternatives to veggie burgers. Algae, when used in conjunction with AI-powered bioreactors, is up to 400 times more efficient than a tree at removing CO_2 from the atmosphere. That means that while we are learning to reduce carbon emissions and augment our consumption patterns, we can start to make big reductions in atmospheric carbon. When wielded correctly, it could make a city carbon negative without changing current production or consumption patterns of the city."

Algae farms are also being developed as a high-protien food source by various companies, including GE, EnerGaia, and iWi.

Algae polymer is being developed by Dutch designers to utlize the raw material to replace plastics, including 3D printing. Another company, Bloom technology makes foam from Algae and uses them currently in shoes and surfboards. At the same time, it cleans the water of pollution and feeds the clean water back to the fresh water ecosystem.

Algae as biofuel is being developed by oil companies such as Exxon and Shell and Venture Capitalists. Such developments can only survive if viewed and supported by governments as means for pollution control and "save the planet."

Mealworms to reduce plastic pollution:
In the December 2019 issue of Environmental Science and Technology journal, Stanford University researchers published their study describing that the yellow mealworms collectively can consume large amounts of plastics and Styrofoam; and excrete only a trace amount of chemicals such as HBCD (Hexa Bromo Cyclo Dodecane). The Environmental Protection Agency from the US and the European Union are concidering a ban on this chemical. HBCD is one of many such chemcials added on to plastics and HBCD is known to cause endocrine disruption and neuro-toxicity.

The mealworms are easy to cultivate and currently used as feed for the farm aniamls and fish that are not harmed by the feed. Humans are also not harmed on consuming such fish, chicken, etc. Anja Malawi Brandon, the lead author of the study says

"The biggest risk is bioaccumulation when it moves up the food chain. What we found was that this flame retardant (HBCD) does not bioaccumulate in the mealworm. That's really exciting. ... It's amazing that mealworms can eat a chemical additive without it building up in their body over time." The article continues to explain that "the mealworms themselves could only eat foams or soft plastics, but scientists believe the bacteria in the guts could in theory also break down harder plastics."

Further research on isolating the bacteria from the gut of the mealworms is needed because potentially the bacteria could break down harder plastics as well. Proactive prevention of use of chemicals such as HBCD is also urgently needed.

3. MIND MANAGEMENT

This book involves mainly this third factor. Mind cannot be completely controlled because that is the nature of mind. However, mind can be managed.

With this understanding, the Chetana Yoga and the techniques of Chetana Yoga for Mind Management are herein described in detail in this book.

Section One:

THE ETERNAL DANCE

THE ETERNAL DANCE

❖ ❖ ❖

Eternal dance is a collection of poems by the author, and is sprinkled throughout the text at the appropriate sections. The book is more or less, a commentary on these poems.

THE ETERNAL DANCE
What a gift of beauty, Nature bestowed
With love, kindness and free-will
Inherent in Life-force, inner nature
What a waste not to hold the heavens
In the chambers of heart
Subtle and eternal, yet within reach
As Awareness
Good intentions
True conscience
Kindness, love, and freedom.
Super-luminal dance, subtle and mystical
With no beginning no end
The eternal dance.

Dr. Rajah Sekaran

AMILEA-MILELA

Amilea, Milela all-pervasive
Life-Energy inherent
In the ocean of Awareness,
We are mere bubbles
In the ocean of Life-Energy
Infinite and eternal,
So, enjoy the life with
The beauty of life, with
Unconditional kindness,
Compassion, love,
Forgiveness and freedom;
No justification, whatsoever,
For frustrations of the mind,
Prejudice of all kinds,
And violence, both direct and indirect,
Isolation and loneliness.

ODE TO CHETANA

When faced with frustrations and
Dilemma of mind
Between right and wrong and
Hundred shades of grey
Causing havoc like a hurricane,
Awaken the heart to Chetana-life
The being aware philosophy,
And a passion
For kindness and compassion
Forgiveness and beauty;
See the Chetana
Of always being aware,
Swallows the dilemma;
Smile and enjoy
The beauty of Chetana.

Dr. Rajah Sekaran

CHETANA YOGA

Chetana yoga creates
A layer of Awareness
Beyond mind, and permanent
Between mind and action
Making room for free-will
To choose the right action,
With good intentions
True to the conscience;
No more commands
From the mischievous mind
In communication and action;
O mind, from now on
I'm not in your beck and call
You have no choice but
To be my friend, forever.

CHETANA – CHAITANYA

Chetana Chaitanya Chit-ananda
Awareness space ākāsha
What is in a name, but carries the
Meaning of a thousand words;
Inherent in life energy
In motion
Spurts of energy
Particles of eternity,
Like fireworks and magic;
Now you see, now you don't
In space limitless, subtle and mystical
Dancing faster than the speed of light
With no beginning, no end,
The eternal dance

MIND IS CLEVER

The mind is very clever
like a chameleon changing colors,
limited by nature, adorns the color
of Consciousness or Awareness
and claims ownership of the Awareness
that is in fact limitless, unchanging, and
Beyond Mind.

ODE TO BEAUTY OF LIFE

You go through childhood and youth
Study and learn all you can
Then starts the real life,
Life of struggles and failures
And frustrations of mind,
Mistakes and missteps
And sprinkles of happiness strewn
Here and there, now and then;
With patience and perseverance
Success and triumphs,
Suddenly you reach the autumn
Of life, in spite of it all,
You look back and see
Acts of kindness, compassion
And Forgiveness on and off,
All through life,
You smile and enjoy
The beauty of life.

PRACTICE CHETANA YOGA

Practice of Chetana Yoga,
The Being Aware philosophy
Emphasized again
Along with the understanding
Every one of us is
A part of the whole
All-pervasive Life-
Awareness space; so,
No justification seen in prejudice,
Isolation and loneliness.

THE TOWN OF CHETANA

Where is the town of Chetana
Founded way back in millenia,
Asked Dennis, the grandson
To grandpa Nelson,
Go look in the Google Earth map
You may find it, we hope,
But you may not find it,
Because we *lost it*,
The town and its people
To the great advancements
In science and technology
That run the world in
Higher and higher gear
That give us comforts, no doubt,
But made us forget
Kindness and forgiveness
Acts of good intentions
With compassion and love,
All expressions of Chetana
That is Awareness unlimited
Being content in the depths of mind
Yet beyond mind
And inherent in life-energy
The very source of happiness and beauty
Chetana, the Awareness unlimited.

Dr. Rajah Sekaran

COBWEBS AND WEEDS OF THE MIND

In one fell-swoop, cut asunder
The cobwebs of the mind
Often in the midst of the weeds
Of mis-steps and past deeds
Tenacious with deep roots,
Using the sword of Chetana yoga
Of *Being Aware*, inherent in *Life Energy*
With the foundation pillars of
Kindness with good intentions,
Compassion and love,
Forgiveness and freedom,
And enjoy the *Peace;*
lest you fall off the cliff of freedom
Into the valley of arrogance and violence
Physical and mental,
And the perils of prejudice,
Practice, practice
Acts of kindness with good intentions,
Compassion and love;
For, the second and third attempts
Become harder and harder,
As the roots of the weeds
Become thicker and thicker.

CRISIS MANAGEMENT

The inspiration for the following poem on 'Crisis Management' was based on a phone call from one who did not believe in the power of Chetana yoga and being aware of the mind. Nevertheless he was desperate due to the thoughts of suicide and homicide; and was seeking immediate help and reluctantly tried my advice, narrated in this poem. He was able to come out of the 'crisis.'

Dr. Rajah Sekaran

CRISIS MANAGEMENT

When waves of crises bombard the mind,
Even thoughts of suicide and homicide,
Pay attention to the chain of thoughts;
Observe continuously again and again.
See the thoughts of the crises
Lose their power and turn powerless
Slowly but surely with Awareness,
Mind then becomes calm and rational,
Find a solution to the root-cause of the crisis.
If it is 'attachment' to a loved one,
'Let Go' the attachment,
For love is not attachment,
Attachment is possessiveness, whereas,
Children and other loved-ones
Are not our possessions.
For attachment leads to 'control and violence'
As mental, physical and/or verbal abuse.
Love is affection and acceptance,
Mutual respect and caring.
If it is events of the past, 'accept',
Accept, whatever it amay be,
For you cannot change the past.
Stop building expectations,
For future is really not in our control.

LOVE AND ATTACHMENT

Mind is from Mars; Love is from Venus.
Attachment arises from Mind, whereas,
Heart Center is the seat of Love and Kindness,
Compassion and Forgiveness.
Attachment is not Love,
Attachment is possessiveness, whereas,
Children and other loved-ones not our possessions.
So 'Accept' them as they are, with Love;
Guidance, of course, everyone needs.
Possessiveness leads to violence
As mental, physical and /or verbal abuse.
So 'let go' attachment and possessiveness;
'Let go' attachment to past events;
'Accept' the past events and future;
Have dreams, ambitions,
Planning and expectations and
Act with full effort, dexterity and commitment;
But 'accept' the outcome; for it is not in our control.
Attachment is from mind,
Love is from Heart, Beyond Mind.

Section Two:

CHETANA YOGA

CHETANA –CHAITANYA

Chetana Chaitanya Chit-ananda
Awareness space ākāsha
What is in a name, but carries the
Meaning of a thousand words;
Inherent in life energy
In motion
Spurts of energy
Particles of eternity,
Like fireworks and magic;
Now you see, now you don't
In space limitless, subtle and mystical
Dancing faster than the speed of light
With no beginning, no end,
The eternal dance

CHAPTER 1

BEYOND MIND

❖ ❖ ❖

*The mind is very clever
like a chameleon changing colors,
limited by nature, adorns the color
of Consciousness or Awareness
and claims ownership of the Awareness
that is in fact limitless, unchanging, and
Beyond Mind.*

BEFORE THE CIVILIZATIONS CAME INTO existence, the primitive way of life was body-focused and instinct based, and the human beings were satisfied with the basic needs of the body, such as procuring food for consumption, processes of elimination, sex for reproduction, and sleep. When civilizations came into existence along with structured societies, the mind was brought into more prominence with the development of knowledge; and different branches of science, art, and literature flourished, rightfully so.

Groupers of knowledge became philosophers and *splitters* of knowledge became scientists. Although advances in science and technology have brought marvelous changes in lifestyle with comforts and longer lifespans, social, economic, and political problems in societies have grown in parallel to the advancements in science and technology. Lately, environmental issues have also multiplied. Use of knife and sword to hunt have been replaced by guns and bombs.

Basic reason, as I see it, is that the human race has become *trapped in its own growth of mind*, being unable to see beyond the individualized mind, somewhat similar to the 'cave parable of Plato' which he used to explain the "Theory of Forms." Plato's Forms are not the forms as we understand in today's world. Plato's Forms are non-physical, non-substance and extra-mental. The Allegory of the Cave of Plato refers to a group of people chained inside a cave which has an opening in the top of the back wall through which sunlight comes through, and the campfire in the back of the cave throws *shadows of people* on the opposite wall. That is all these chained cave dwellers of Plato have believed to be the real world. However, when one of them escapes, he finds that the real world is outside the cave. When he returns and describes what he saw beyond the cave, his fellow captors considered him to be insane.

Religion is beautiful in promoting devotion and love which can be the best alternative to the present-day mindset of violence and prejudice. Unfortunately, however, the religious and racial differences and prejudice have become the root-cause of violence and intolerance globally, based on racism and religion on many occasions, both in the past and present.

So many religions came into existence but could not solve the basic cause, mentioned above. The individualized mind has become as much *trapped in its own evolutionary growth*, as science and technology advanced; it is now showing an increased use of drugs, drug-related violence, prejudice-related violence, violence related to personal issues and mental dysfunction and problems of sexual harassment, as an *escape mechanism*.

Spirituality as distinct from religion also has tried to answer the problem for thousands of years through the *mind and intellect* with marginal success. Perhaps because it could not reach the understanding of most common people; spirituality remained esoteric and elusive.

Milela theory described in this book is finally addressing the problem by proving *Awareness is inherent in Life Force itself and is beyond mind*; and *Chetana Yoga* is described as the *method* to live such an Awareness based life.

This is my humble attempt to explain *what is Beyond Mind*, to see if such a *shift in Paradigm* will help to alleviate these problems in the individual level, as well as the problems in the society, politics, and environment that are plaguing this modern society all over the world. I see this as the *natural next evolutionary growth* of humankind to mature on, from *body-mind focused life to Life-Force-Awareness focused life*. That requires a clear understanding of what Life Force-Awareness is and what mind is.

MIND AND THE THEORY OF SPACE AND SUBTLETY

The theory of continuous and relative Subtlety, to be described in one of the following chapters, divides the *space* arbitrarily into,

1. the least subtle space
2. the subtle space and
3. the most subtle space.

So, in the context of the Space and the Theory of Relative Subtlety, the individual mind belongs to the *Least Subtle space* a.k.a. the gross space. The mind is described as *subtle (the least subtle)* because it is *intangible and exists both in the body and outside the body*.

The Universal unlimited mind belongs to the *Subtle* world or *Subtle* space.

The Limitless Awareness is indeed the *Most-Subtle* that is inherent in the all-pervasive Life Force or Life Energy which remains as a *potential Force* in the

Limitless Awareness with no manifestation of forms in the *Most-Subtle* state; and manifests as *Gravity and Electro-Magnetism* which pervade *the subtle and the least subtle states.*

1. THE LEAST SUBTLE (GROSS) SPACE AND THE MIND:

For all practical purposes, mind in the least subtle or gross space, is further divided into:

1. Mind that collects data from the sensory functions such as sight, hearing, touch, taste and smell. The brain responds to the data by sending out motor commands appropriately to the corresponding parts of the body.
2. Intelligence is responsible for analyzing, evaluating, inquiring and deliberating conclusions. It is also the site of ego personality.
3. Memory: Thoughts and ideas are stored as memories.
4. Ego: Provides the personality of the person based on his or her self-impression of the past achievements, thoughts, memories and the future plans and ambitions.

All four are important in a person's personality and for normal functioning. Mind itself is not the villain. It is the animal nature of anger, jealousy, undue craving, and arrogance that *attach* themselves to the personality that causes havoc for the person and for the others. Whereas in acts of *unconditional kindness, love, forgiveness and good intentions*, in thought, communications, and action, the first thing that drops off is the *attachment to* ego along with pretentiousness and unwanted pride associated with the negative qualities listed above. Then one by one, all the negative qualities from anger to jealousy drop off.

2. THE SUBTLE SPACE AND THE MIND

The Universal mind is beyond the individual mind and is considered to belong to the Subtle Space. The individual mind is limited by the individual ego and personality, even though the individual mind is still a part of the Universal mind. Until now, all the philosophers, physicists, and thinkers have considered the Universal mind same as Limitless Awareness or Consciousness. However, *Awareness is beyond the individual mind as well as the Universal mind*. Nevertheless, the qualities of the Universal Mind lead one to the realization of *Limitless Awareness*, by dropping off the *limited* nature of the individuality of the mind.

The Universal mind includes *compassionate qualities of kindness, forgiveness and compassion*. These qualities are beyond the ego-centered mind; and the *attachment* to ego drops off automatically, in the practice of the compassionate qualities mentioned above.

It is reasonable to assume that the *compassionate qualities* of the subtle universal mind, are associated with the *heart*, in a *subtle* way. That is why we often say, a person is *'heart-broken'*, to describe the state of mind when some tragedy occurs in his or her life. Moreover, *Broken heart syndrome or Takotsuba cardiomyopathy* is a recognized entity. It was described as a multivessel spasm by Sato and others. Two thirds of the patients have a preceding stressful event such as a death of a loved one or a major accident.

In 'Broken heart syndrome', the patients present with the same symptoms as the heart attack (ACS– acute coronary syndrome).

However, coronary angiogram shows no obstruction; and left ventriculography shows *apical ballooning* of the heart, a characteristic of this syndrome. It is a treatable heart failure. It is more frequent in women (90%).

In our own experience, qualities of the subtle mind such as compassion and love are often felt in the heart.

Forgiveness is another quality of the subtle mind. Especially in the understanding of forgiveness, all the great teachers of the past have emphasized *forgiving oneself first with the good intention of not repeating the same error ever*, as essential to be able to forgive others. Then only, the mind will be truly ready to forgive others.

3. AWARENESS-LIFE FORCE AS THE MOST-SUBTLE SPACE

In the context of Relative Subtlety again, Awareness-Life Force is of the *Most-Subtle* space in which the Life Energy remains as a *potential Force*. Awareness-Life Force includes the life energy itself that is in all things in Nature, in both animate like humans, animals, and plants, and inanimate objects like mountains and rivers, in other words, the whole Universe that includes billions of galaxies.

Moreover, the Life Force is inclusive of Awareness, that is the inherent intelligence in the Life Force. Being *limitless*, it is not affected by the problems of the individual mind that is always *limited* by its ego.

Limitlessness is inherent in the Life Force-Awareness. So, being all-pervasive, all things are included in the Life Force that includes the individual mind.

Unconditional *compassion*, naturally is the expression of Limitless Awareness-Life Force. Perhaps, this expression of *compassion*, forgiveness, *kindness*, and *good intentions* also will give an answer to the enormous problems created by the individual mind.

WHAT IS MIND

Mind, is the individualized facet of the universal mind; and is a facilitator for the functions of the brain, at the individual level, as a component of the psycho-neuro-endocrine system of the body. Similar to the thalamus in the brain, the mind acts as a relay station to collect the data, such as sensory data and initiates appropriate action by the brain. If brain is the anatomical part, the individualized mind is the physiological component of the nervous system. Nevertheless, mind is both inside the brain and outside the brain; and as the universal mind, it is also outside the body. To the extent, the mind shows the qualities of the universal mind, such as kindness, good intentions, compassion and forgiveness, the mind reveals its original state of the universal mind.

John Donne, a seventeenth century philosopher, was right, when he said,
No man is an island, entire of itself

Mind and its Divisions
Moreover, mind is also described as

    A. Individual Mind
    B. Universal Mind
    C. Collective Mind.

The description of the mind in the least subtle (gross) space is nothing but an *individualized universal mind*. In other words, mind by nature can be situated in any place, in time and space, either near or far, far away, unlike the other functions of the brain that are localized to the body itself. That is why mind is always *subtle and universal*, but it appears to be individualized due to the application of the functions of the mind pertaining to the body itself. For example, in the *feedback* mechanism of the psycho-neuro-endocrine system that maintains the *Internal Milieu* of the whole body, the mind part of the psycho-neuro-endocrine system is mainly involved in the functions of the body. Similarly, the emotions of the mind that is from the *limbic* system of the brain is an individualized function of the mind. In the functioning of the mind as a collector of sensory data, memory and as the ego-personality again, the mind is certainly individualized.

Whereas, when the mind is planning a vacation for example, it can be far, far away from the body and cross many continents in a moment's notice. In this context, the mind is of the *subtle space* travelling in time and space, way beyond the limitations of the *Least Subtle (Gross) space*.

The *collective mind* is not universal mind; and definitely not the universal Awareness. We need to understand the difference.

The collective mind can be the most productive or the most dangerous collection of individual minds depending on their *basic theme* of being collected. It can be very constructive. On the other hand, it can harbor arrogance, prejudice, vanity, jealousy and anger. This tendency of collection of minds unfortunately seems to be prevalent in different forms in different societies all over the world. When the negative thoughts listed above, drop off due to the practice of *kindness and good intentions* in thought, communications and actions, the same collective mind is no different from the Universal mind that abides in the Limitless Awareness.

CHAPTER 2

CHETANA YOGA — *THE NEXT STEP IN HUMAN EVOLUTION*

❖ ❖ ❖

*Chetana yoga creates
A layer of Awareness
Beyond mind, and permanent
Between mind and action
Making room for free-will
To choose the right action,
With good intentions
True to the conscience;
No more commands
From the mischievous mind
In communication and action;
O mind, from now on
I'm not in your beck and call
You have no choice but
To be my friend, forever.*

EVOLUTION IS DYNAMIC AND NOT static. As the next step in the human evolution in this 21st century, herein is described the mind-training technique to realize Awareness and to live an *Awareness-based life rather than a mind-based life. Chetana Yoga, in essence*, creates a permanent layer of Awareness *instantaneously*, between the mind and action, making room for the *free-will and focus*. Thus, the free-will is strengthened to choose the *right action–either right communication or right action–*; and the person is not obligated to do *everything* the mind commands. The mind, mischievous and temperamental at times, can be a *friend or foe;* and the person has the power to direct the mind to be a *friend always, automatically*.

MIND-TRAINING TO REALIZE AWARENESS

Considering the above discussions in the previous chapter–*Beyond Mind–*, some training is needed for the mind through education, to practice the limitless Awareness-Life Force, on a daily basis effortlessly.

To realize the limitless Awareness through *meditation* is possible. However, it is very complex and time consuming. So herein, I will describe a *technique* based on the *Milela Theory* (briefly described below first, mainly for the understanding of this technique.) The technique can be easily practiced by anybody in the normal daily life, without much effort. I call this technique, *CHETANA YOGA*, or *CHAITANYA YOGA*. *Chetana* or *Chaitanya* in Sanskrit means *Awareness*. The method of Chetana Yoga follows a brief description of the Milela Theory.

I AM MILELA–THE THEORY EXPLAINED
Milela is the acronym for *Mutually Inherent Life Energy and Limitless Awareness* which is described in detail, in the chapter–The Milela Theory.

Rene Descartes (1596-1650) a philosopher, mathematician and scientist asked over 300 years ago, "I know that I exist; the question is, What is this 'I' that I know (that I exist)"

The question is appropriate. However, the answer, "I think; therefore, I am/exist" he came up with, has been proven wrong by the advances in Science in the 20th century. Discoveries in the Biological Sciences showed that there is life in plant kingdom and even in *bacteria* in the micro-biological world. The *Quantum theory* came along to show that the basic unit of life and all matter is the *atoms* with their nuclei with protons and neutrons, and the electrons swirling around the nuclei at a far distance from the nuclei, with huge empty spaces, at a tremendous speed nearly the speed of light (186,000 miles per second). Other subatomic particles also play their role in the origin, sustenance, annihilation and rebirth of all things in nature.

This technique of Chetana Yoga depends on the realization that the *Inner Nature* (also known as *Swarupa*) of all things in Nature is the *limitless Awareness-space* with *Inherent Life Force (Milela)*. Similarly, as individuals our *Inner Nature* is also the *Limitless Awareness Inherent in the Life Energy (Milela)*. Mind and body are merely instruments within the *Limitless Awareness (Inherent in Life Force)*, which is known as *Chetana or Chaitanya* in the languages of Sanskrit and Tamil.

The Three Steps of the Practice of Chetana Yoga

In the practice of Chetana Yoga, paying attention to the mind is more important than paying attention to the body and the surrounding.

1. Chetana Yoga of the surrounding
2. Chetana Yoga of body
3. Chetana Yoga of mind

1. Chetana Yoga of the Surrounding

In the method of the Chetana yoga of the surrounding, start being aware of the sounds, you hear, such as the cars, metro train with their typical sound of the train on its tracks and siren on and off and the natural sounds of the birds and squirrels, moving around the trees; then generally seeing the surrounding, whatever it may be, and feeling the gentle breeze on the skin etc,. In other words, using the senses of hearing, seeing, smelling and feeling the breeze on the skin etc., to go beyond the senses. Similarly, we see different people surrounding us, with acceptance of the differences. We are not meant to be clones of the same body and mind.

2. CHETANA or CHAITANYA YOGA of BODY:

By being Aware of the actions of the body, both in collecting data through the sense organs and in the motor actions, Chetana Yoga establishes the person and the personality in Limitless Awareness. The sense organs include the eyes, ears, nose, tongue and the skin performing the functions of seeing, hearing, smelling, tasting and touching respectively. The motor actions include the actions of upper and lower extremities, speech and reproduction and elimination.

For example, being Aware of the movements of yoga and especially the breathing, during the practice of yoga poses and pranayama, brings about the real nature i.e., the Limitless Awareness, that is always Beyond Mind. I call this Chetana Yoga or Chaitanya Yoga of the Body.

The understanding of Chetana Yoga of mind and body by everyone, or by a majority of the population, may not be possible in the beginning of the Paradigm shift in this evolution of the humankind. However, if this understanding takes root by way of education in schools and colleges and in leadership camps that are now becoming more common in the corporate and political world to reach the leaders in education, environment and politics, it can bring about lasting changes to shed off and shrug off prejudices born out of racism, religion, caste systems and class systems that have led to violence in this country as well as in the rest of the world.

Principles and steps to follow for the success of Chetana Yoga are further dealt with in the following chapter–Epiphany.

3. CHETANA YOGA OF MIND

Mind is subtle, dynamic and powerful; it can remain as calm as the surface of a lake or can become exuberant, as the surfing waves of the ocean on a full moon night, or act up as a dynamite, as we have witnessed in the minds of the mass-shooters in the recent years. So, the mind needs to be handled with utmost care. Chetana Yoga is herein described to do just that.

Chetana or Chaitanya Yoga is mainly based on being Aware of the mental processing, as and when it occurs; and this eventually becomes a habit. The thoughts or sensory data received can be seen just like one can see the hand in front of the face. Notice the thoughts as a flow, one thought after another and another, like a chain. Being aware of the input to the mind, whatever it may be, gives ample opportunity to act, with kindness and good intentions to oneself, as well as to others. Seeing the hand is by using the eyes; and seeing the mental processing, as they occur in daily life, is done, by just being Aware of the mental process, as it occurs. In other words, just like you are aware of seeing your hand moving or lifting an object, you can be aware of your mind and feelings; and eventually it becomes a habit and a part of living.

In being aware of the mind and the mental process, the paying attention or being aware is not through the mind but is through the *heart center*, the seat of kindness, forgiveness, and compassion. This "being aware" is the ultimate reality and it is true in all the faiths and all the paths one may follow. If one follows the path of devotion (Bhakti Yoga), or the path of action, with kindess, good intentions, and compassion (Karma Yoga), or the path of wisdom of knowing that the ultimate reality is the Limitless Awareness from the heart center (Gnana Yoga), they all have to end in the heart center being the seat of the Limitless Awareness that is not limited by the mind and its problems of prejudice, arrogance, anger, vanity, and fear.

Chetana Yoga, by being Awareness-based, creates a permanent layer between the mind and the action; and gives an opportunity to evaluate the action instantaneously, before it happens. Thus, it strengthens the Free-Will to choose the right action–either right communication or right action with good intentions–and keeps the mind as a life-long friend.

Mind and its components of both negative qualities and positive qualities do not disappear, just like the hand does not disappear in the Awareness, when you are aware of the hand. However, all your mental processing, including receiving of sensory data and thinking and the actions become Awareness based, and not mind based or instinct-based; giving an opportunity for the free-will to select the right action, not swayed by the idiosyncrasies of the mind anymore, such as arrogance, greed, vanity, prejudices and violence; and the negative qualities get corrected and neutralized as a part of living. This training is based on the new Paradigm of 'The Milela Theory' that I am introducing in this book and that dictates the Life Force-Potential and the limitless Awareness are inherent of each other and beyond mind; and this Awareness-based life is the natural progression of human evolution. However, when such Awareness is confined to the mind, as only a concept, it is no longer limitless. When practiced as being Aware of the mental process itself, such Awareness is limitless and not within the control of the individual mind and its frustrations. It is the mind that cannot be otherwise straightened out similar to the old saying 'you cannot straighten out a dog's tail.'

There is no doubt that mind can reach Awareness in the deeper realms of the mind, in deep meditation, because the Awareness is the basic substratum of all things in the universe and beyond, including the mind. However, when the mind comes out of the meditation, the real nature of the mind takes over that includes the ego, memory, and other preconceived ideas of prejudice and negative qualities of anger, anxiety, jealousy, and arrogance. The mind is no longer established in Limitless Awareness; the mind remains limited within the confines of the limitations of the negative qualities mentioned above. Chetana Yoga releases the mind from such limitations and bondage.

MINDFULNESS AND CHETANA YOGA— A CLARIFICATION OF THE DIFFERENCE:

Mindfulness has become popular now and is also commercialized as mindfulness-shirts, mindfulness-pants, and mindfulness-yoga mats etc. Various techniques of mindfulness are also published in journals. Mayo Clinic for example describes the technique of using the senses of seeing, hearing, touching, etc. of your surroundings and development of "Paying Attention." For example, enjoy the food you are eating and 'live in the moment.' Next it moves on to paying attention to the breathing and body described as the 'Body Scan Meditation', which means, lying down flat and paying attention to each part of the body. Some meditations also involve paying attention to thoughts during meditation. In this mindful meditation, paying attention to the mind is done during the meditation only and the mind is back to the same set of problems of anxiety, loneliness, depression etc.

Chetana Yoga takes you one step further to pay attention permanently to the mind itself and to the flow of thoughts as and when they occur. It does take some practice and eventually becomes effortless. It creates a permanent filter or a layer between the thoughts and action, especially reactions, for example in moments of anger and anxiety. It is further explained also in the Chapter on Mind Management.

In this process it is important to accept yourself and accept the mind and the flow of thoughts as a part of the body, same as accepting the hand, for example, as a part of the body. When the hand goes near a flame, it is withdrawn automatically. Similarly, paying attention to the flow of thoughts as and when they occur directs the person to the right action, instead of reacting to thoughts without a filter.

Mindful of Mind

Mindfulness or Awareness is a more fundamental reality than mind, because Awareness, commonly referred to as Mindfulness, is a fundamental reality behind mind and body, atoms and subatomic particles, galaxies and planets, in short, all things in nature.

Crisis Management

The inspiration for the following poem on 'Crisis Management' was based on a phone call from one who did not believe in the power of Chetana yoga and being aware of the mind. Nevertheless he was desperate due to the thoughts of suicide and homicide; and was seeking immediate help and reluctantly tried my advice, narrated in this poem. He was able to come out of the 'crisis.'

Dr. Rajah Sekaran

CRISIS MANAGEMENT

When waves of crises bombard the mind,
Even thoughts of suicide and homicide,
Pay attention to the chain of thoughts;
Observe continuously again and again.
See the thoughts of the crises
Lose their power and turn powerless
Slowly but surely with Awareness,
Mind then becomes calm and rational,
Find a solution to the root-cause of the crisis.
If it is 'attachment' to a loved one,
'Let Go' the attachment,
For love is not attachment,
Attachment is possessiveness, whereas,
Children and other loved-ones
Are not our possessions.
For attachment leads to 'control and violence'
As mental, physical and/or verbal abuse.
Love is affection and acceptance,
Mutual respect and caring.
If it is events of the past, 'accept',
Accept, whatever it amay be,
For you cannot change the past.
Stop building expectations,
For future is really not in our control.

ODE TO CHETANA

When faced with frustrations and
Dilemma of mind
Between right and wrong and
Hundred shades of grey
Causing havoc like a hurricane,
Awaken the heart to Chetana-
Awareness and a passion
For kindness and compassion
Forgiveness and beauty;
See the Chetana
Of always being aware,
Swallowing the dilemma;
Smile and enjoy
The beauty of Chetana.

Section Three:

EPIPHANY

CHAPTER 1

THE EPIPHANY

"I believe in God, but not as One thing, not as an old man in the sky. I believe that what people call God is something in all of us. I believe that what Jesus and Mohammed and Buddha and all the rest said was right. It is just that the translations have gone wrong"

— JOHN LENNON (1940-1980)

THE WORD 'EPIPHANY' IS DEFINED in the Merriam Webster Dictionary as "a usually sudden manifestation or perception of the essential nature or meaning of something."

Epiphany One

I am herein describing an *epiphany*, I had, that describes the relationship of the mind and its ego to *Awareness*.

Awareness or Consciousness is not a part of the mind, as it has been generally conceived. *Awareness is inherent in the Life Force itself*, both in the individual level and in the Universe, at large. It is all-pervasive in the Universe in both the animate such as human beings i.e., all of us, and the inanimate. Being mutually inherent, Awareness and Life-Force are one and the same and is the True-Self or *True 'I'*.

Awareness or Consciousness, being all-pervasive is also in the mind. However, when Awareness is expressed or experienced by the mind, it takes on the *limited* nature of the mind. But Awareness-Life Force, the very life of all things in nature, is *limitless*. Such Limitlessness can never be contained in the *limited mind*. However, when the mind sheds the *limitations* of prejudice, arrogance and violence, the mind experiences and expresses the *Limitlessness* of Awareness inherent in the Life Force, for example, in the acts of *compassion and kindness with good intentions*.

Limitless Awareness being all-pervasive exists in all individuals. In reality however, *we all exist in such Life Force-Awareness*. So, every individual has the potential to live a life from such Life Force-Awareness perspective. However, we are all needlessly limiting ourselves, by living a life from the mind's perspective with built-in prejudices of all types.

Epiphany Two

Until now, Spirituality has been teaching Awareness or Consciousness for thousands of years *as a part of the mind*. The reality is, mind by nature is always *limited* within its confines of collecting data, memory, personality and intelligence, and by nature, acts within a *limited extent*. Whereas Awareness is *limitless* and is the very basis of mind; and Awareness is *inherent in the Life-Energy, that is keeping us alive*, and remains as the substratum of all things in Nature, as explained in the theory of *Milela*. Spirituality, nevertheless, *did not dissociate the limitless Awareness or Consciousness from the mind* which forever enabled ignorance and is unable to realize the *Limitlessness* of Awareness-Life Force. The mind being limited in nature was only able to understand the Limitlessness as a *concept or knowledge* unable to *realize* the Limitlessness to the full extent, beyond the concept. So, this realization remained elusive.

The *epiphany* of this author is to realize that Awareness is not a part of the mind; and spirituality has been teaching Awareness or Consciousness as a part of the mind and did not dis-associate Awareness which is *limitless* from mind which by nature is *limited*. This prevented the mind from fully realizing and practicing Awareness on a continuing basis which is now made possible by *Chetana Yoga*.

Awareness is neither a part of the individual mind nor the Universal mind. Universal mind, in relation to the individual mind could be unlimited, but not limitless as Awareness is.

Moreover, in the context of the Theory of Relative Subtlety described by this author in the Chapter on Theory of Relative Subtlety, the individual mind belongs to *the Least Subtle state or space*. The Universal unlimited mind belongs to *the Subtle state or Subtle space*; and the limitless Awareness belongs to the *Most-Subtle state or space*.

To conclude, the Awareness is beyond mind, whereas mind arises from Awareness and mind depends on It for its existence.

Discussion

We have had many spiritual masters who came and taught the values of divine qualities of love, compassion, kindness and forgiveness and described the *mind as the problem* that creates sufferings in life. Nevertheless, their teachings were misinterpreted and misdirected to the mind to manage their teachings.

People continue to suffer from anger, anxiety born out of fear, from prejudice and violence in mind and its actions. The reason is that mind claimed ownership of

Awareness in spite of the mind being dependent on the Awareness for its own existence. However, looking for an answer to solve the problems of the mind, through the mind, is like having the tricky fox to guard the sheep herd. Awareness and Life Force are inherent of each other and do not depend on the mind.

For example, Gautama Buddha came 2,500 years ago and taught Awareness as *Chetana*, in the language of *Pali*, a dialect of Sanskrit; and *Pali* was prevalent at that time as the common people's language in Kapilavastu, prince Siddhartha's kingdom in the foot-hills of Himalayas. He also said that the *mind is the root cause of all the sufferings of the mankind* after he struggled through and realized the Ultimate Truth. However, in due course *Chetana* became associated with mind. When it came to a translation to English, *Chetana* was interpreted as *mindfulness* in the deeper realms of the mind, rather than calling the *Chetana* as *pure Awareness* itself. So, it led to the wrong direction to realize Awareness through the mind, leading to more frustrations. Whereas, *Being Aware* of the mental process itself, including thoughts and ideas, will lead one to practice Awareness, on a *continuing basis.*

Mind is only an *instrument* similar to body. Mind is called the *antah-karanam* in Sanskrit, meaning *inner instrument*, the body being the *outer instrument*—both body and mind, being instruments, nevertheless.

SIX PRINCIPLES FOR THE UNDERSTANDING AND SUCCESS OF THE CHETANA YOGA:
Understanding the following principles are necessary for the success of *Chetana Yoga*, described in the previous chapter.

1. Mind is in *Awareness*. Mind depends on Awareness-Life Force for its existence.
2. *Awareness-Inherent in Life Force* exists independent of the mind— in waking, dream sleep (or REM-rapid eye movement sleep) and deep sleep states.
3. Awareness inherent in Life Force as the *source of the mind*, is sometimes misunderstood as the *conscience* (mana-sakshi in the languages of Sanskrit and Tamil meaning 'witnessing the mind'). Awareness and conscience (or mana-Sakshi) are not the same. Awareness that is the *Most-Subtle* is the source of conscience as well as the mind. Conscience is *Subtle* and is the connection between Awareness and the mind.
4. Forgiveness, love and compassion, kindness, and good intentions *true to the conscience* emanate from the Awareness Inherent in Life Force in a *subtle* way. These qualities have a subtle connection, also with the *heart*.
5. Kindness and good intentions true to the conscience together form the key to realization

of Awareness and *freedom from sufferings of the mind.*
6. Awareness-Inherent in Life Force is *all-pervasive;* and all things in Nature and the Universe exist in the Life Force Inherent in Awareness.

STEPS FOR FREEDOM FROM SUFFERING

1. Body exercises such as yoga, pranayama, and meditation are useful in taming the body and mind.
2. Chetana Yoga or Chaitanya Yoga:
3. Being *aware* of the mental process, including thoughts and ideas of the mind, *as they occur,* on a continuing basis, objectifies the mind as an instrument and tames the mind.
4. In the practice of Chetana Yoga, *kindness and good intentions* originate from the mind at the heart level, in a subtle way. Such kindness and good intentions *true to the conscience,* then naturally are carried on to all communications including speech, and actions. In such actions of kindness and good intentions true to the conscience, the *attachment* to the ego falls off; the true 'I' and *the personality* of the

individual *remain*, now devoid of anger, arrogance, greed, vanity, anxiety born out of fear and jealousy.
5. *Forgiveness* is an essential basis for the practice of compassion, kindness and good intentions to prevail in the Chetana Yoga of being aware of the mind. All the great spiritual teachers have stressed on the importance of forgiveness of others as well as oneself. Only when a person is able to forgive oneself, and to truly resolve not to repeat such an action ever, he or she will be capable of forgiving others.
6. Ignorance is part of the ego as 'I-thought'. When ignorance goes away at the dawn of knowledge, *that the reality of Awareness-Inherent Life Force is the basis of body and mind*, the true 'I' and the ego as the personality *remain*, now devoid of ignorance of the *Reality*.

QUESTIONS AND DISCUSSION:
1. One may ask, "What about soul?"

Gautama Buddha in his great teachings answered this question by saying *'Anatha'* or *'Anathma'*, meaning there is no such thing as individualized soul or selfhood.

According to the *Milela Theory*, the Life Energy Inherent in Awareness is considered to be the *Universal Soul*, just for the sake of understanding. When it is viewed as individualized, it is viewed as *the individual soul*.

So, the so-called Individual Soul is actually the Universal Soul or the Life Energy Inherent in Limitless Awareness that is individually seen as the soul by the *mind*. From the point of view of Universal Soul however, there is only Universal Soul; and mind and body function within this Universal Soul which is the substratum.

2. One may ask, "What is the connection between the so-called Universal Soul and the body and its mind?"

The connection is subtle. The so-called Universal Soul is actually the Mutually-Inherent Awareness-Life Energy. We will call it, *It*, for convenience. At the *brain level*, 'It' connects at the intelligence part of the mind.

The ego part of the mind, from the *brain level*, can be the source of anger, anxiety, jealousy, etc.. However, at the *heart-level* 'It' connects as kindness, forgiveness, love and compassion, conscience and good intentions true to the conscience etc., all divine qualities. That is why people make a distinction between 'living from the head level' or 'living from the heart level.'

3. One may ask, "Is there a rebirth or reincarnation?"

Gautama Buddha was asked by a scholar 2,500 years ago, "you say, *Anatthama* meaning there is no individual soul and you also believe in reincarnation. How is it possible without the individual soul traveling to the next birth?"

Buddha answered: "Yes, the cycle of birth and death is true. For example, planting the seed from a tree goes on to form the same type of tree."

Now we know it is due to the genetic code in the seed that forms the same type of tree or plant. Similarly, humans are born again and again based on the *fertilized ovum* with the DNA code that determines the characteristics and features of the newborn.

That is why children inherit not only some of the physical characteristics of their parents, but also some of the mindset of the parents.

However, it does not explain why children are not the same prototype of the parents or between the children of the same parents. The children are *not the exact clones* of the parents. Very often there is a stark contrast between the siblings as well as between the parents and their children, especially in their mindset and personality. Even though the brain is an anatomical part of the body, the mind itself is *subtle and exists both inside the body and outside the body*. To reconcile with our *Theory of Subtlety*, explained under the chapter, 'Space and Theory of continuous and relative Subtlety', it is reasonable to assume that the mindset and personality of the newborn all through his/her life depends also on the previous birth as well as the environment. It is based on the actions from the previous birth as well as unfulfilled *strong* desires with *attachment* and *cravings* of the mind. They are called

Vasanas in Sanskrit and *MunVinai* in Tamil. However, in the previous birth, as well as in the present birth, if a person forgives himself or herself and truly resolves not to repeat such wrong actions ever, they are not carried forwards to result in *Karma-Palan*— fruits of actions— in this life as well as the next life, if there is one, as explained below.

It goes without saying that if one follows the *New Paradigm* of Chetana Yoga or Chaitanya Yoga, with the right *attitude development* recommended as the steps to follow for success (*Siddhi*) in Chaitanya or Chetana Yoga, there is no more rebirth; no more sufferings of the mind and body, as a result of repeated birth and death cycle; and that is *Freedom*. In this birth also, it gives clarity and there is freedom from frustrations and sufferings of the mind

AWARENESS AND MIND:
Awareness is Universal and Limitless whereas mind is individualized and limited by nature.

Adi Shankara lived in the 9th century between 788 and 820 C.E. and is known for reviving the Advaita Vedanta philosophy or non-dualism, throughout India. Advaita teaches Jeevathma (the indivual soul) and Paramathma (the Supreme soul or Limitless Consciousness or Brahman) are one and the same. Adi Shankara's domination in Indian philosophy and Vedanta continued on for many centuries. When

Vedanta was imported to the West in the 20th century, the philosophy of non-dualism and the knowledge of limitless Consciousness were introduced, especially after Vivekananda (1863-1902) came to the Congress of all Religions in Chicago in 1893 C.E. Consciousness and Awareness are synonyms and carry the same meaning. However, Consciousness was considered as a component of the mind, in the Philosophy of both East and West, leading to much confusion.

The mind was considered as the ultimate reality, in the West, mainly due to the teaching of René Descartes (1596-1650 C.E.) philosopher, mathematician, with his famous quote *"cogito, ergo sum" in Latin*, meaning "I think, therefore I am/exist" that led to the concept that mind and body are separate and it placed the mind at a superior position. This Cartesian thinking continued on until the 20th century, when mind and body are proven to be one unit, mainly based on the description of the *Internal Milieu* by Claude Bernard and the discovery of multiple hormones and neurotransmitters that connect the body to the nervous system through a network of 100 billion neurons and trillions of synapses and neurotransmitters.

Claude Bernard, (1813-1878 CE) a Physiologist from France, described the constancy of the internal environment and coined the word *internal milieu* in the humans in 1854, long before the discovery of hormones and neurotransmitters. However, it took another 50 years before it was fully recognized in experimental Physiology. Originally, he described the *internal milieu* based on temperature control and blood sugar control as the basis of *homeostasis*. The insulin hormone that controls the blood sugar level, however, was discovered much later in 1921 by Frederick Banting, a physician and Charles Best, a medical student from University of Toronto. More discoveries of multiple hormones, enzymes and neurotransmitters in the 20th century give a more complete picture of the constancy of the *internal milieu* and also the body-mind connection and the feedback mechanism. *The Hypothalamic-Pituitary axis* is an example of the *feedback* mechanism of multiple hormones.

Psycho-neuro-immuno-endocrine system evolved, and the feedback mechanism regulating the *internal milieu* further strengthened the case for the integration of body and mind as one unit. Further, the psycho-neuro endocrine system along with the immune system showed the interrelationship between the different systems of the body. In fact, all the systems of the body and mind are interrelated. For example, when the metabolism of the body is slowed, the message is relayed through the network of neurons and neurotransmitters to the *Hypothalamus* which secretes the TSH-Releasing factor (thyroid stimulating hormone-releasing factor.) Hypothalamus is responsible for releasing multiple releasing factors for various hormones. TSH-Releasing factor then acts on the *Pituitary* gland to secrete more TSH (thyroid stimulating hormone) which increases the secretion of thyroid hormone from thyroid gland which brings the metabolism up. On reaching an optimal level the Hypothalamic-Pituitary axis once again regulates the *thyroxine* secretion from the Thyroid gland to an optimal level and maintains the *constancy of the metabolism*. Likewise, the interrelationship of the various systems of the body and mind are continually regulated through all the hormones, enzymes, and neurotransmitters to maintain the *internal milieu*.

In a broad sense, it is not too far-fetched to say that all systems in the Universe are interrelated and interdependent to maintain the constancy of the environment. However, this balance in Nature, being what it is, can be disturbed by the actions of human beings; and in alleviating the imbalance, the constancy of the environment needs to be maintained for their optimal maintenance.

Proof of Awareness being Beyond Mind

Mind cannot contain the limitless Awareness and definitely cannot be the source of Awareness that is all-pervasive and limitless. That is proven by the example, when a person is in deep sleep or in fainting state, the mind loses the consciousness and is turned off, similar to a computer being turned off.

Here again, the Awareness that always exists *does not ever become nonexistent* in those instances of fainting or deep sleep. When the person wakes up, the mind and its functions, including memory, come back to life because the body is kept alive by the Life Force-limitless Awareness which exists both inside the body and outside the body.

GET OUT OF THE ZOO

By the word 'zoo', what this author is trying to convey is the individualized mind that is filled with anger, jealousy, vanity and violence and trapped with prejudices and unscrupulous, unfulfilled and sometimes unethical cravings. So, there is no room for kindness or good intentions in thought, communications and actions. The mind is trapped or imprisoned, similar to the *cave parable of Plato* in which Plato uses the allegory to explain "The Theory of Forms." So, herein the trapping of the mind is similar to the 'cave parable' until the Awareness is realized as being *Beyond Mind*.

The individualized mind is a component of our bodily functions and needs to be recognized as a part of the physiology of the body and neither separate from the body nor superior to the body.

How can any individual go *Beyond Mind* and live a life that is beyond the problems of the mind? There is a technique to objectify the mind, as described under the heading 'Mind-Training to Realize Awareness', as the technique of *Chetana Yoga* or *Chaitanya Yoga*, described in the previous chapter.

From the *theological point of view* also, for the sake of freedom from the cycle of birth and death associated with all the sufferings of the life with occasional spurts of *happiness*, one needs to practice *Limitless Awareness that is Mutually-Inherent in Life Force*, as described in the *Chetana Yoga* or *Chaitanya Yoga* in the previous chapter.

This *Being Aware* is not only limitless, but it also relieves the frustrations of the mind, such as anger and anxiety and loneliness or feeling of isolation from friends and family.

CHAPTER 2

LIMITATIONS AND LIMITLESSNESS

Knock, And He will open the door
Vanish, And He will make you shine like the sun
Fall, And He will raise you to the heavens
Become nothing, And He will turn you into everything

— RUMI

THIRU MOOLAR, A *MYSTIC AND poet*, who lived 3,000 years ago in South India wrote an invaluable collection of spiritual teachings, known as *ThiruMantiram* which has 3,000 verses. One of the verses, written in Tamil, is given here.

மரத்தை மறைத்தது மாமத யானை
மரத்தில் மறைந்தது மாமத யானை
பரத்தை மறைத்தது பார்முதற் பூதம்
பரத்தில் மறைந்தது பார்முதற் பூதம்.

Marathai Maraithathu Māmatha Yanai
Marathil Marainthathu Māmatha Yanai
Parathai Maraithathu Parmuthal bootham
Parathil Marainthathu Parmuthal bootham.

THE TRANSLATION:
When the form (and name) of wooden, wild elephant is seen, the *wood* (as the basis) is hidden (not seen.)

When the *wood* is seen (as the basis of all forms made of wood), the different forms and names are hidden (not seen.)

When All Things in the world are seen, the *Param* (as the basis) is hidden (not seen–not realized.)

When the *Param* (the highest or supreme reality as the basis of all things in the world) is seen (realized), the different forms and names are not seen. – end of translation.

In other words, to reconcile with the *Milela theory* introduced in this book, the racial and other differences and prejudices between people are not seen, when the highest reality of Limitless Awareness is realized and practiced.

The word *Param* is common to both Tamil and Sanskrit languages. Param is used here as 'Parathai and Parathil' and is translated as supreme, highest of the highest and beyond.

For our discussions, *that* which is supreme and beyond all things, is also *beyond mind*. All things in the world are only *limitations* of the *Limitless Awareness* that is beyond mind, similar to *space*. Space is always *space* both inside and outside of an object whether it is enclosed in a box, house, or galaxy. Similarly, Limitless Awareness is the same whether it is in a limited form of an African, Afro-American, Asian or Caucasian body (listed in alphabetical order). It is the same in rich or poor, man or woman.

The languages that developed in the early period in India, thousands of years ago, are now referred to as Indo-European languages in the British Encyclopedia. Sanskrit was spoken and written in languages of different regions in the script of that particular region.

When Sanskrit was used in South India where Tamil was the language, it was written in Tamil script; and is now known as *Grantham*. Moreover, both Tamil and Sanskrit are now declared as the two *classical languages of India* by the government of India.

In both Tamil and Sanskrit, *Param* means the same thing. In the Sanskrit-English dictionary by V.S. Apte as well as in Tamil-English translations, *Param* means beyond, supreme, and highest of the highest.

Even though *Limitlessness* is the basic substratum of all existence, limitation is the reason for the existence of all things in Nature. If Awareness remained limitless, it is nothing more than a *Nothing*. This *Nothingness* is what was named by Buddha as *Sunya*, which means zero or nothing. Because starting from the subatomic particles to atoms to molecules to matter and multiple billions of galaxies, *limitation* of the *limitless* is what brought them into existence. However, the *Limitlessness or Nothingness* continues on with the *limitations* as the subatomic particles, *dancing* in the Limitless Awareness Space with inherent Life-Energy.

The Most-Subtle space–Life Energy Inherent in Awareness–became lesser and lesser in *subtlety*. For example, it is similar to *space* itself which takes up different forms and names depending on whether it is a room, a box, a planet, or a galaxy which are nothing but space enclosed.

The very formation of atoms with nucleus from the subatomic particles of neutron, proton, electrons was a process of limitation from Limitlessness.

What Are Other Limitations?

There are also social, political, economic and environmental limitations that need to be accepted and balanced and corrected within the context of limitlessness.

Social problems of racism, caste systems and religious differences are prime examples of limitations in the society at large. Denial of such differences and ignorance of all these limitations, *being within the context of limitlessness*, is what makes such social differences pathological with formation of *hate groups* in different parts of the world.

Similar are the political and economic problems that need to be addressed and corrected, knowing the limitation of all these limitations *being within the context of limitlessness.*

The common thread in all these issues seems to be *ignorance* that is *missing the forest for the trees*, as the saying goes, which is trying to explain how on focusing on the trees, for example, the interrelationship between the trees and between the trees and the forest is ignored. Similarly, ignorance of the limitlessness promotes problems of the limitations of the society overall.

Environmental limitations and imbalances are mostly human-made and need to be corrected by education and practical solutions. Implementation, however is becoming increasingly difficult. So, understanding the problems within the context of limitlessness, at least by some of the leaders in education and politics will help the realization of our responsibility to the future generations.

So, all of a sudden, we cannot go from all these problems of limited outlook to limitlessness as the answer. Besides, the broader concept of limitlessness appears to be beyond the scope of the limitation itself; whereas limitation and limited outlooks are within the scope of the *limitlessness*. In fact, all the various systems of the society are dependent on Limitless Awareness-Life Energy as the very substratum, knowingly or unknowingly.

So, the paradigm-shift, necessarily has to be a shift from the problems of limited and narrow outlook to the *Limitless Awareness-Life Energy*-based-life. Then, individuals and societies and countries can live peacefully with *acceptance of the limitations* within the Context of the *Limitless Awareness-Life Energy*. This shift in *Paradigm* is not only recommended but is essential for the *ultimate survival of humanity*. Limited outlook that gives rise to problems of social, economic, and political differences is the breeding ground of prejudices. Besides the understanding aspect of this shift in Paradigm, there is inherent power in limitless Awareness-Life Energy that will help with the solutions of the problems of prejudices borne out of such limited outlook. Limitations are part of life and the world, and as such cannot be eliminated. However, such problems borne out of racial, religious differences and imbalances in economy including the widening gap between rich and poor and men and women, can be resolved by realization of the *new Paradigm shift*, that is the *Limitless Awareness-Life Force being Beyond Mind*, encompassing all the systems of the world.

There is no denying that the mind is an essential part of our life and survival. However, the Limitless Awareness-Life Energy is *the basis of mind* and needs to be realized as such, for intelligent and permanent changes to happen in the imbalances and prejudices in the mind, and the resultant problems, discussed above.

Section Four:

THE MILELA THEORY

CHAPTER 1

THE MILELA THEORY

❖ ❖ ❖

Mutual Inherence of
Life Energy and
Limitless Awareness

Science is not only compatible
with Spirituality; it is a profound
source of Spirituality.

— CARL SAGAN

IN ONE OF THE TWO great Epics of India, Mahabharata (the other being Ramayana), King Yudishtara, the eldest brother of the five Pandavas in exile, was asked many questions by a lake-demon. That was the demon's condition to bring his four younger brothers back to life, after the brothers swooned, on drinking the lake water. The King answered them correctly and saved his brothers. One of the questions was "what is the most surprising thing in the world?" The King answered; "people see others dying and taken to the funeral pyre every day and yet they think they will live forever." The demon was satisfied with that answer and the brothers of King Yudishtara came back to life.

Besides the two epics of India, there are thousands of such stories some of which, I presume were added on subsequently. When I was growing up in India, I used to think these ancient stories are a waste of time, when we were asked to sit in gatherings, where these stories were narrated by a speaker. There was of course no TV in those days. However, I enjoyed the *sweet* (usually, sweet Pongal or Kesari), at the end of the discourse. But later in my life, I realized how cleverly our ancestors have interwoven the knowledge of science, medicine, astronomy and social reform into these stories passed on from generation to generation. Some stories of social reform included devotion to parents and spouse, respect and devotion to teachers and even speaking against caste prejudices.

Applying a similar concept today, derived from the story of King Yudishtara narrated above, we see babies born all the time and people passing away every day; yet we take the *Life Energy Principle*— that *appears* to come and go with the body— for granted. We ignore the most *fundamental* of this very existence, the Life Energy. We also ignore *'Awareness'*, that is *Inherent* in the Life Energy, even though we are aware of the Awareness. Instead, we confuse the *Limitless Awareness* with the limited states of consciousness and subconsciousness, that are part of the mind.

> The mind is very clever
> like a *chameleon* changing colors,
> *limited* by nature, adorns the color
> of Consciousness or Awareness
> and claims ownership of the Awareness
> that is in fact *limitless*, unchanging, and
> Beyond Mind.

THE MILELA THEORY

MILELA is an acronym for *Mutually Inherent Life Energy and Limitless Awareness.*

From this point on, the expansion of *MILELA Theory* namely the *Mutually Inherent Life Energy and Limitless Awareness* will not be repeated; and it will simply be referred to as *Milela Theory*, for the sake of convenience.

The concept of Life Energy being Universal has been practiced by ancient civilizations of Greece and India. In the Greek mythology, Atlas, from the Classical art point of view, supports the whole globe and the celestial spheres, in other words, the whole Universe. Gaia was described as the god of Earth. Poseidon of the Greek mythology and Neptune of

Roman mythology carrying the trident represent the oceans.

In the mythology of India, Booma Devi is described as the goddess of the earth and all that is on earth. Varuna is the god of the oceans and is responsible for the rainfall.

The Ganga and Kaveri Rivers are worshipped as gods. All the Life Energy together is also worshipped under various names, such as Mother Durga, MahaKali, ParaShakti and Parvati. However, the philosophy behind all the mythology of India insist on One Supreme Being as the Absolute Truth. The Native Americans, mistakenly called American Indians, worshipped earth and all the forces of Nature as God. In fact, their whole lives are closely tied to Nature.

More recently, Gaia Theory was introduced by James Lovelock, a chemist and further reinforced by Lynn Margulis, a microbiologist in the 1970s as Gaia Hypothesis and now slowly being more and more recognized by the Scientific community. It advocates that the Earth has life just like we human beings have life.

Milela Theory is presented here as a *General Theory of all things*. Moreover, the Milela Theory is introduced to bring a paradigm shift and wants to bring a realization that the Life Energy is Universal and exists not only in the earth but also in the whole Universe. According to this theory, *Life Energy is Mutually*

Inherent in the Limitless Awareness. The Life-Energy inherent in the pure Awareness is all-pervasive and is not only the basis of mind in individuals but is *Beyond Mind* and is the basic substratum of all things in Nature.

Be being beyond mind, the *Milela Theory* is attempting to cure the problems of the mind plaguing the whole world with violence, racial, regional, and caste prejudices, mistrust and wars.

It is a well-established fact that the human body has an *internal milieu* which was introduced by Claude Bernard in the 19th century in the year 1854. The internal milieu or the internal environment is maintained by a *feedback mechanism*. In the 20th century, the discovery of many hormones and neurotransmitters further reinforced the understanding of the feedback mechanism. Similar to the maintenance of the *internal milieu* with the feedback adjustments with the psycho-neuro-hormonal system of the human body, the earth and its components have an internal milieu that balances itself. Gaia Hypothesis was the pioneer in explaining that the earth has *life* and as one of the evidences, the Gaia Hypothesis used a simple example and pointed out the exchange of oxygen and carbon dioxide, being opposite in the plants versus the animals and human beings. Namely, the plants take in CO_2 and release O_2. Human beings and animals take in O_2 and release CO_2.

Milela Theory is now proposing that the Life Energy Inherent in Awareness-Intelligence is the reality, maintaining the environment with adjustments similar to the feedback mechanism of the *internal milieu* of the human body; and these balancing adjustments in the environment sometimes bring about hurricanes, eruption of volcanoes, earthquakes, and floods and of course, normal rains. These adjustments, nature does make and will make to bring about a *balance* in its *internal milieu of the environment.* So, environmental damage by the actions of humans does produce global warming, floods, etc. until balance is achieved.

In summary, Milela Theory also proposes that there is an *internal milieu* of the world as well as the Universe with feedback mechanism similar to the *internal milieu of the body*, making adjustments to bring about a balance for the whole world and the Universe.

It is a well-known scientific fact that only four percent of the Universe is known to human knowledge at present. The remaining 96% of the Universe is still not well understood. This 96% includes the *Dark Matter (24%) and the Dark Energy (72%).* It is also well-accepted in Physics that the atoms and subatomic particles such as *electrons and bosons* form the basic unit of all things in the Universe. Moreover, Milela-theory is inclusive of both the 4% of the Universe that is known, and the 96% the Universe that is not yet clearly understood.

MILELA AND THE TRIPLE PRINCIPLE

Milela functions through the *Triple Principle*, described here.

The *Triple Principle* consists of

1. Desire or need or teleological purpose
2. Knowledge and
3. Action

Every *action* needs to have a *desire or need or a teleological purpose* to initiate such action. Action also needs a certain *knowledge*, similar to DNA of genetic code, to carry out the *action*. The Triple Principle applies universally to animate and inanimate, humans and animals, plants and insects, bacteria and viruses, air and fire, water and earth. It is hard for our *limited mind* to comprehend the Universal nature of this Triple Principle beyond our limited body and mind. With unlimited mind, that exists both inside and outside the body, it is possible to understand this Triple-Principle. Viewing from a *supra-mental* approach from Milela point of view, it is possible to understand. From Chetana or Chaitanya Yoga point of view also, it is possible to understand. Chetana, as explained earlier, is the *limitless Awareness* that is all-pervasive; and *Chetana or Chaitanya yoga described earlier,* explaining how to be aware of the thoughts and actions, as and when they occur, *is recommended as the next step in the human evolution.*

The Triple Principle of desire, knowledge and action is the expression of *Life Energy* itself. The same Life Energy of Milela along with the Inherent Awareness Space possesses the Triple Principle of desire, knowledge and action to give rise to *Gravity space* and *Electromagnetism* to begin with. Again, the Triple Principle of Milela forms the very basis of *space and time* from *Gravity* and *Electromagnetism*. The manifestation of all the subatomic particles resulting from the attraction and repulsive forces of *Gravity and Electromagnetism*, resulting in a tremendous movement nearly the speed of light, leads to the formation of matter by the *interaction* of *Electro-magnetism, and Weak and Strong Nuclear Forces*, as proven by the physicists.

So, it is no surprise, the Triple Principle of desire, knowledge and action, of *Milela* is the basis of all systems, both animate such as human beings, animals, plants and insects, and inanimate such as air, fire, water, earth, oceans, rivers and mountains which are projections of the earth on the earth.

One may ask how DNA-like action of the *knowledge of the Triple Principle* is possible at the early stages, even before atoms come into existence. The atom is first formed with electron, proton and neutron, and the atoms then form molecules leading to elements of carbon, hydrogen, nitrogen, oxygen, sulfur and phosphorous, which are all essential for the formation of *life*. The elements enumerated above lead to the formation of DNA of the genetic code. So, it is only reasonable to ask the question–how does DNA-like action come into existence even before the formation of atoms and genes in the beginning–?

According to the Theory of Subtlety mentioned earlier and described fully in the next chapter, *only the least subtle, a.k.a. the gross space*, is bound by time and space as experienced by us. *The Most-Subtle space is beyond the concept of space and time; and the Life Energy inherent in Awareness in the Most-Subtle* provides the *DNA-like* action as the very source of all that exists in the world and Universe, within time and space. The Most-Subtle *Life Energy* inherent in Awareness provides the DNA-like action, somewhat similar to the *virus that contains DNA*, in the process of *endo-symbiosis and symbio-genesis* in bacteria and other living systems, in the subsequent evolution, hypothesized by Lynn Margulis.

So, once again, the discussion above proves the Life Energy Inherent in Awareness Intelligence is *all-pervasive* and provides the energy for existence of all things in Universe.

It is mind-boggling to comprehend how subatomic particles manifest as billions of galaxies and how all *matter* appears, disappears and reappears constantly. It is perfectly reasonable to infer that the immensity and precision of the workings of the *cosmos* can only function and exist with a high degree of intelligence–the *Limitless Awareness*.

For every action some degree of intelligence is needed to provide the *knowledge* of the act. So, such intelligence has to be present in the *Life-Energy* that initiates the action, similar to *DNA and the genetic code* of our world.

The *knowledge* and the *desire or need* of the Triple Principle provide examples of the similarity to DNA and the genetic code as we know it. Such *knowledge* of the Triple Principle similar to DNA is not only the basis of Origin of all things in Nature but also for their maintenance starting from Gravity and Electromagnetism, space and time, air, fire, water and earth, and all things in the world as well as in all the galaxies.

Moreover, the DNA-like code of the *knowledge* directs the *action* based on the *desire or need* which also originates from the DNA-like code.

Similarly, the *energy particles forming all matter* have to have an *Inherent Intelligence* to provide the knowledge to form matter.

I call this Intelligence as the *Inherent-Awareness-Intelligence*; and the energy particles are from the *Life-Energy* that is all-pervasive. This Awareness-Intelligence is inherent in the all-pervasive Life-Energy; and this all-pervasive Life-Energy is inherent in its Awareness-Intelligence. So, they are both *mutually inherent and inseparable* from each other and *non-dual*. I call this the *Mutual Inherence of the Life-Energy and Limitless Awareness—MILELA* as an acronym.

Nondual Milela and Duality

When the nondual Life-Energy inherent in Awareness-Space differentiates into *Gravity Space and Magnetism*–which later was named Electro-Magnetism by Maxwell's Electromagnetic theory–the duality was born. This duality continues on as Dark matter and Dark Energy and as space-time duality.

Then the Gravity and Electro-Magnetism interact to form the subatomic particles. The particle-wave duality of the subatomic particles, nucleogenesis, formation of atoms and molecules are followed by differentiation to air, fire water and earth and billions of galaxies, stars and planets and Energy-mass duality. The subatomic particles are characterized by a tremendous speed of movement which varies among different particles, nearly close to the speed of light. As the speed of the movement of the subatomic particles varies for the different particles, different forms come into existence based on the number of protons in the nucleus of the atoms, and molecules.

However, the speed of movement for each particle, for example electron, is the same for all electrons.

In the following discussion on *Life-Energy and Awareness*, even though they are both inherent of each other, as *MILELA*, Life-Energy and Awareness are discussed separately as follows.

CHAPTER 2

LIFE ENERGY OF MILELA

❖ ❖ ❖

LIFE ENERGY–INHERENT IN AWARENESS
LIFE ENERGY IS THE VERY basis of self; and Life Energy individualized is the self. Life Energy-Force is *love, freedom, beauty and happiness.*

LIFE ENERGY IS LOVE
Life Energy is the basis of *love; Life-Energy is love;* everyone loves being alive. When a loved one is snatched away by separation or death, the *bonding between the two lives* suffers and it feels like the end of the world. When a child is pulled away from the mother, the cruelty of such action causes a great agony for both the mother and child. The unconditional love between the mother and child suffers.

Life Energy is Beauty

ODE TO BEAUTY OF LIFE

You go through childhood and youth
Study and learn all you can
Then starts the real life,
Life of struggles and failures
And frustrations of mind,
Mistakes and missteps
And sprinkles of happiness strewn
Here and there, now and then;
With patience and perseverance
Success and triumphs,
Suddenly you reach the autumn
Of life, in spite of it all,
You look back and see
Acts of kindness, compassion
And Forgiveness on and off,
All through life,
You smile and enjoy
The beauty of life.

When we realize the Life Energy as the *reality* that is behind all that we perceive, and all that is within our Awareness, we see *beauty*. This reality as well as *beauty* is further appreciated in acts of kindness, forgiveness and compassion.

LIFE ENERGY IS FREEDOM

For example, *freedom* to live a life of one's choosing is such a fundamental right; when such a choice is taken away by oppression of a population, people are even willing to die for the sake of freedom. Moreover, freedom to think, speak and write is as fundamental as life itself. That is why Life Energy is freedom.

LIFE ENERGY IS HAPPINESS

Life Energy inherent in Limitless Awareness takes us beyond the idea of the mind being the source of Awareness. The mind-based life we are living in, has given moments of happiness on and off and often associated with an equal amount of suffering somehow, somewhere, sometime. Happiness and suffering duality of the mind is the nature of mind, because mind comes under the confines of the *Law of Opposites*. The law of opposites says the *duality* of all things in Nature reflect the law of opposites and can only survive with maintenance of a *balance* in their manifestation.

Whereas, the Awareness mutually inherent in Life Energy is beyond the limitations of the mind and beyond the law of opposites. The spurts of happiness come and go in the activities of the mind; whereas the happiness is the very nature of Awareness inherent in Life Energy; hence, the happiness is ever present.

AWARENESS AND LIFE ENERGY

Human beings *confuse* Awareness as a part of the mind, whereas Awareness is beyond mind and the proof for the same is given, in the next sub-chapter on Limitless Awareness.

Similarly, we *confuse* the presence of Life Energy with the five *senses* of hearing, seeing, touching, tasting, smelling and also *emotions*. Whereas Life Energy is all-pervasive and present even in the absence of *senses*, for example in the atoms with protons and neutrons in their nuclei and electrons circling them at a tremendous speed nearly the speed of light, giving rise to mass, further leading to formation of molecules and structures, both animate and inanimate. In the macroscopic level, Life-Energy is present, when the senses are absent in *deep sleep*.

All things in the Universe have *structure and function* similar to anatomy and physiology of the human body as well as that of mammals and plants. In the *structure*, atoms are considered the basic unit; and the *function* of rapid movement and interactions of the subatomic particles result in the formation of nuclei, atoms, cells and molecules, leading to *macroscopic* forms of earth, water, fire, air and humans, mammals, plants, and insects on this earth. They also form *microscopic* forms such as *bacteria and viruses*.

The microscopic structures of bacteria and viruses also perform an important role in the *various functions* of the earth. It is common knowledge that different bacteria and insects produce *compost* from waste products. They also produce the basic elements of life such as carbon, oxygen, nitrogen, sulfur and phosphorous in the earth.

We now know the bacteria and viruses can cause diseases, thanks to Louis Pasteur. However, we have not even scratched the surface of the *knowledge* to unravel the secrets of the various life-saving *functions* performed by the bacteria and viruses. We are just beginning to understand the gut bacteria of human beings and mammals in the production of certain proteins and vitamins in their intestine. Some of the vitamins produced by the gut bacteria include biotin and vitamin K and some of the hormones, one of which regulates the storage of fat.

The *mitochondria*, considered to be the *battery*, providing *energy* for the functions of the plant cells are, in fact, a bacterium that was planted (pun not intended) into the plant cell about 700 million to 1.5 billion years ago, as per the theory described in the late 20th century by Lynn Margulis, as an argument for *symbiosis, symbio-genesis and endosymbiosis* forming new forms of life. Lynn Margulis also speculated that early forms of life came into existence by endosymbiosis of viruses with their DNA into bacteria. Lynn Margulis, a microbiologist, theorized *Gaia Theory*, along with James Lovelock, a chemist.

The very existence of life on Earth for plants, animals and human beings depends on sunlight. For example, in the plant life, photosynthesis depends on sunlight. For animals and human beings, sunlight is an essential energy needed for survival. There would be no rain without sunlight. So, energy in different forms is essential for life on Earth. The definition of life, proposed in the 20th century, included respiration and elimination of waste products as a requirement. If we follow such definition, energy such as sunlight will not qualify. Even some of the bacteria and all viruses will not qualify. So, a Paradigm Shift in the definition of life is needed for the 21st century.

So far, up to and including the Gaia system, all the definitions of life have used the cell as the basic unit and its characteristics of structure and function as the basis of life.

What is Life?

The definition of life has been evolving for many centuries, based on the understanding of the scientific community in each period, as the knowledge of the wonders of Nature have been uncovered and unfolded.

Starting from the idea of *life* being confined to humans and animals the understanding has been more and more inclusive, as the knowledge and discoveries have been changing. The definition then included plants, insects and bacteria.

Structure and function, similar to anatomy and physiology of human body were then considered as essential to life. Dissipative structures and *ecosystems* were described in the late 20th century to explain the food cycle and the exchange of oxygen and carbon dioxide and the metabolism and the waste products which are broken down by bacteria and insects into carbon, oxygen, nitrogen, sulfur and phosphorous and recycled in the *ecosystem*, to form new life.

René Descartes (1596-1650), in the seventeenth century, said "I think, therefore I am." In the philosophy of Descartes, the thinking function of human mind was equated with *existence*. If we split his *principle in*to two, namely "I think" and "I exist", existence per se obviously includes life in general; but existence should include inanimate also. So, in the interpretation of the Cartesian Principle we need to assume that the Cartesian Principle equates existence and life. To the extent all existence is life, the Cartesian idea of "I am" or "I exist" can be acceptable. However, in the interpretation of "I think," the first part of the Cartesian Principle, the mind *'to think'*, as the criterion of existence has been disproven in the 20th century, by proving that there is life, not only in humans and animals and plants which are all *eukaryotes* with nucleated cells, but also in bacteria which are *prokaryotes*, meaning non-nucleated, but containing DNA and RNA.

With the introduction of Gaia Theory by James Lovelock and Lynn Margulis in the 20th century, the *whole Earth* is now understood as *a living system*, but extends only up to the atmosphere above and a few miles beneath the surface of the Earth. Gaia Theory compared it to the outer layer of a tree being alive and over 90% of the trunk of the tree being dead wood.

The 20th century definition of *dissipative structure as the basis of life*, meaning respiration and elimination of waste products, is now open to question, when we consider that there are some bacteria and all viruses do not fit the definition.

Viruses are nothing more than DNA and RNA. Similar to bacteria, some of the viruses cause disease by invading the host and releasing the DNA by breaking down its own protein membrane. DNA then uses its energy to replicate in the host. However, the role of viruses in the beginning and maintenance of life process is still not well understood. Perhaps the viruses also play a role in the maintenance of *internal milieu* of individuals as well as that of the world and the Universe, similar to the present-day understanding of the role of bacteria in maintaining the *internal milieu*.

Similarly, the seed of a plant has DNA and the genetic code which has the potential to produce life.

So, DNA appears to be a form of *energy* and appears to be the very basis of life in the Universe. Viruses that contain DNA can be also defined as a form of *energy* which is activated inside the host. But we cannot ignore the fact, all forms and all molecules, including DNA, are made of *atoms* with their sub-atomic particles as the basic unit of life.

Besides, Modern Physics has taught us that the atoms are the basic unit of all matter. The subatomic particles are a form of *energy*, also giving the *appearance* of mass due to their tremendous speed of movement, *at nearly the speed of light*. It was proven by the Special Theory of Relativity of Albert Einstein (1879-1955), $E=Mc^2$, where c^2 represents the *constant* of the *speed of light (186,000 miles per second) squared*, and M stands for *mass*, and E stands for *energy. Energy and mass are different forms of the same thing.*

So, from the discussion above, *energy* seems to be the common denominator of *all that exists* and *all that is life*. So, considering the presence and manifestations of Energy in all that exists and in all that is Life, *Energy or Life Energy- is the new definition of Life.*

Considering all the understanding we now know presently from Modern Physics, Biology, Microbiology and Biochemistry, *Energy* is not only the common denominator but the very basis of *all that exists* and *all that is life*. So, it is *Life Energy; Life-Energy is life*. This Life Energy is mutually inherent in Awareness according to the *Milela Theory*.

LIFE-ENERGY OF MILELA AND MODERN PHYSICS

Modern Physics was born in 1900 CE, when Max Planck, considered as the *father of Quantum Mechanics*, discovered that the subatomic particles behave as *quanta* or packets of particles rather than as waves. It was contrary to what was shown unequivocally by Thomas Young in1803 with the *double-slit experiment*. His experiment showed that the light travels as waves. The controversy lived on until it was shown that the subatomic particles behave as both particles and waves by Louis de Broglie. In his thesis, he also said "the greater the momentum of a particle, the shorter the corresponding wave-length". That is why matter waves are not visible in our macroscopic world; moreover, the subatomic particles are very, very small.

Physicists discovered that all light waves including radio-waves travel at the constant speed of 186,000 miles per second. Planck also described the Planck's constant—constant of proportionality—that higher the frequency of light quantum, higher is the energy.

The next mile-stone was the description of the *Special theory of Relativity* by Albert Einstein in 1905, as $E=Mc^2$, and showed the kinetic energy has mass, and mass is another form of energy and c is the constant of the speed of light. In 1915, Special Theory was followed by Einstein's *General Theory of Relativity* that proved the *warping of the space-time continuum*.

In the same year of 1905, Einstein also published the theory of *photo-electric effect*, for which he received the Nobel Prize in 1921. His theory of photo-electric effect was based on the experiments by Philippe Lenard, for which he received Nobel Prize in 1905 showing the electron was released immediately, when light hit the metal. Einstein's theory explained that the electron particle was knocked off the atom from the metal piece, when it was hit by a light particle, photon; and also proved the quantum nature of light energy particle.

Even though the quantum theory was found to be *correct* by many different experiments in the microscopic quantum world, it was described to be *incomplete* by many physicists, including Einstein. In the macroscopic world also, the *commonsense* ideas of ordinary events were found to be *incomplete*.

In 1964, Bell's theorem was published by J.S.Bell, a physicist at CERN (European Organization for Nuclear Research) in Switzerland with mathematical proof. Since then it had been explained with multiple different examples. Gary Zukav, being a non-physicist, has clearly summarized the essence of the Bell's theorem, in the language that can be understood by non-physicists, in his book, titled "The Dancing Wu Li Masters", including some of the comments of the theorem by some of the advanced physicists, showing the importance of the Bell's theorem in Physics and the unexplainable *connectedness of quantum phenomena*, as follows:

To explain the Bell's theorem, Polarization of photons in pairs shooting out in opposite directions, *but identical with each other*, was created by electrically exciting a gas that emits light and sending them through polarizers. It is similar to the EPR (Einstein-Podolsky-Rosen) thought experiment in which *spin states* were used instead of states of polarization. In the Bell's theorem, the polarization was made to be *vertical or horizontal*. Behind each polarizer *in the opposite direction*, photo-multiplier tube was placed to make a click, each time it detects a photon.

When both polarizers are similar–either vertical or horizontal–the photons (going in opposite directions), emit the clicks *same time*. If one of the polarizers was made to be opposite of the other–e.g., one vertical and the other horizontal–a click on one side was not accompanied by a click on the other side. In all different combinations, there was always a *correlation in spite of being separated in space*. This correlation can be predicted by the quantum theory.

When *local cause* was suggested to explain the *correlation*, as commonsense should, the local cause was disproved by Clauser-Freedman experiment. The experiment was further improved later, in 1982, by Alain Aspect, a physicist from University of Paris, France, verifying the statistical predictions of *quantum mechanics* and leading to the phenomenon of *superluminal transformation of information*.

Super-luminal communication (faster than the speed of light) transfer of information was once again suggested, as originally speculated by Einstein-Podolsky-Rosen experiment.

From Psychology point of view, Carl Jung's (1875-1961) *concept of synchronicity* also suggested *a wholeness behind the connectedness of parts of unconnected events.* In the words of Henry Stapp (1970), "The important thing about Bell's theorem is that it puts the dilemma posed by quantum phenomena clearly into the realm of microscopic phenomena… (it) shows that our ordinary ideas about the world are somehow profoundly deficient even on the macroscopic level."

Henry Stapp further wrote in 1977, "Quantum phenomena provides *prima facie* evidence that information gets around in ways that do not conform to classical ideas. Thus, the idea that information is transferred superluminally is, a *priori*, not unreasonable.

…Everything we know about nature is in accord with the idea that the fundamental process of Nature lies outside space-time… but generates events that can be located in space-time. The theorem of this paper supports this view of Nature by showing that superluminal transfer of information is necessary, barring certain alternatives… that seem less reasonable. Indeed, the reasonable philosophical position of Bohr seems to lead to the rejection of the other possibilities, and hence, by inference, to the conclusion that superluminal transfer of information is necessary."

David Bohm wrote in 1974, (Parts) "are seen to be in immediate connection, in which the dynamical relationships depend, in an irreducible way, on the fate of the whole system (and, indeed, on that of broader systems in which they are contained, extending ultimately and in principle to the entire universe). Thus, one is led to a new notion of unbroken wholeness which denies the classical idea of analyzability into separately and independently existent parts..."

To reconcile the Theory of *Milela* with the present-day knowledge of Modern Physics, the *Forces of Gravity Space and Electromagnetism* are derived from the *wholeness of the Most-Subtle Space of Mutually Inherent Energy and Awareness* in which Life Energy remains as a *potential*. This also further leads us to the understanding that the *space and time* are derived from *Interactions* of *Gravity-Space and Electromagnetism*, the 'offspring' of the *Mutually Inherent Energy and Awareness*. There is no time and space (and all things within the space), without this *Mutually Inherent Energy and Awareness-Space* and its offspring, Gravity-Space and Electromagnetism that, according to quantum mechanics, lead to the formation of trillions of subatomic particles. Then the formation of atoms, molecules, cells and different galaxies, stars and planets follow.

LIFE ENERGY AT THE LEAST SUBTLE (OR GROSS) SPACE

It is common knowledge that humans, animals, and insects have life. Plants also possess the same Life Energy. This was proven by Jagadish Chandra Bose in 1901, using his invention of an instrument called Crescograph. It has multiple gears and a smoked glass plate that records the movement of a plant's tip under a magnetic scale of 1/10,000. The movement of the tip of the plant is shown in the plate. He presented this experiment in Royal Society London, England in May of 1901. He also proved that plants have a reproductive system and have Awareness of their surroundings. We all know how plants and trees bend towards the direction of the Sunlight, even if it takes a convoluted course.

Life Energy as the definition of life as discussed in the previous pages, is everywhere. In that sense, inanimate, such as mountains, oceans, and rivers also have *life* because they all contain *atoms* which are a form of *energy*. Mountains have a much longer lifespan than the humans, animals and plants. Mountains also show movement even though the movements are of a much smaller scale. For example, the Heart Mountain is a mountain range of about 100 kilometers long and it is known to have moved 100 kilometers about 50 million years ago in the area of the present-day border between Montana and Wyoming. Movements of mountains on a much smaller scale and in the present-day are also known to occur. According to the Theories of Plate Tectonics and Continental Drift, the mountains as well as the continents with oceans are known to be slowly moving. They are all fundamentally atoms and subatomic particles which have Life Energy and movement of tremendous speed nearly the speed of light, at the subatomic level.

Life-Energy in our world, is what keeps the inanimate and animate, including mammals and we, the humans alive. When Life Energy leaves the body or plants, the body or the plant is said *to be dead*. The body is declared to be dead because of the stopping of the vital systems, such as the cardio-vascular system, nervous system and pulmonary system and the kidneys in the body. This stopping of function takes place at varying threshold levels for the various systems of the body. For example, when the systolic blood pressure falls below 90 mms of mercury, the glomerular filtration of kidneys stops; kidney stops functioning; but it is reversible, when the systolic blood pressure goes above 90 mms of mercury. However, if the pressure goes to zero and remains there, the vital functions of heart, lungs and brain as well as kidneys will fail, leading to death.

The Life Energy that only *appears* to leave the body at the time of death, remains at the Subtle state and the Most-Subtle state at *deeper dimensions of the space*. In fact, Life-Energy does not leave; it is the body and its functions that stop functioning completely, at the time of death. Due to the dis-association of the *subtle* Life Energy from the body (or the plant), the dead body (or the plant) disintegrates to *nitrogen* etc. until it is used up for the manifestation of plants and other living things again, at the Least Subtle Space (a.k.a. Gross space).

Changes in the Least Subtle-Gross Space is the norm. It is said that the only thing that is permanent in the world is *change* itself. So, the changes to the body, such as disease, aging and death happen to the body due to the *dis-association* of the *Life-Energy-Awareness* from the body either *gradually or suddenly*. This *individualized* life-Energy Awareness always remains in the *Universal Life-Energy-Awareness* at the subtle space and the *Most-Subtle* space. So, the changes including the diseases, aging and death happen *only* to the *body*; however, from the mind point of view, it appears as though, the Life-force is leaving the body at the time of death, whereas in fact the changes, and dissolution and eventual disintegration occur in the *body only* and not in the Life-Energy force, which is the *Most-Subtle wholeness* that is not subjected to the changes of the parts.

LIFE-ENERGY AT THE SUBTLE SPACE

To begin with, it is *Life Energy that is Inherent in Limitless Awareness* that gives rise to Gravity and Electromagnetism. Gravity by nature has *attractive force*. It is called Ākarshana *Shakti* in Sanskrit. Whereas Electromagnetism, called *Kānta Shakti* in Sanskrit, has both *attractive force and repulsive force;* and has electrical charge of positive, negative and neutral. The *opposite* charges–of negative charge and positive charge–show *attractive force*. The *similar* kind of charges–either both positive or both negative–shows *repulsive force. Gravity* and *Electromagnetism* by nature have *movement*. These opposing forces of attraction and repulsion interact and create a movement of tremendous speed that results in *appearance* of trillions and trillions of subatomic particles such as *bosons, electrons, protons, neutrons, and positrons* which are nothing but Life Energy in different forms. The subatomic particles are either negatively charged or positively charged or neutral. For example, the electrons are negatively charged and the protons are positively charged and the neutrons are neutral.

The subatomic particles also interact with each other in all possible, multiple ways transforming to billions of new particles continuously in this *dance of the Quantum world*. In both the Subtle space and the Least-Subtle space, Gravity-space provides the *stage for the dance.* The opposing forces of attraction and repulsion of Electro-Magnetism creates a movement of tremendous speed, resulting in formation of trillions of subatomic particles, similar to the churning of milk to produce butter.

In the description of *The Big Bang Theory* in Physics, electrons and positrons (known as antiparticle) destroyed each other in trillions. If such destruction was a complete *annihilation*, atoms would not have formed and there would not have been a formation of the Universe with billions and billions of galaxies including our own *Milky Way galaxy*, we live in. However, in the early annihilation, the electrons prevailed in their encounters with the positrons. Moreover, the electron and positron annihilation release two *photons* shooting out at the speed of light continuously for each electron and positron. In our world of the Least-Subtle space, the electrons orbiting around the nucleus, which contains the protons and neutrons, together form the atoms, the basic unit of all things. The electrons are also known to release energy, as they jump close to the nucleus from an excited position away from the nucleus.

SOME OF THE PARTICLES SUMMARIZED
Electrons are negatively charged elementary particles described as fermion. They were discovered in 1897

by J.J. Thomson. They are known to interact with *Gravity*, *Electromagnetism*, and *Weak Nuclear Force*. The anti-particle is positron.

Positrons are positively charged and were discovered by Carl Anderson in 1932, four years after it was theorized by Paul Dirac. Positrons are also fermionic similar to electrons. The anti-particle is electron. If the positron-electron annihilation is at low energy, it produces gamma ray photons. They are used in PET scans (Positron Emission Tomography) in Nuclear Medicine.

Protons are positively charged elementary particles that were first observed by Eugene Goldstein in 1886 and was later named by Ernest Rutherford in 1917. One isotope of the Hydrogen atom is a single proton. The number of protons distinguishes each atom from the next. Protons and neutrons are bound together by the nuclear force to form the nucleon, forming the nucleus of the atom.

Neutrons have no net charge and they were discovered by James Chadwick in 1932, twelve years after it was theorized by Rutherford. Along with protons, they are called nucleons and they form the nucleus.

Bosons are particles named after Satyendra Nath Bose, who described along with Albert Einstein, the Bose-Einstein statistics. There are two classes of particles, one being bosons, and the other being fermions (all of the particles described above the 'bosons' are fermions). Bosons follow the Bose-Einstein statistics. Recently, Peter Higgs and Francois Englert were given a Nobel Prize in 2013 for independently discovering boson particles.

Photon is one of the elementary particles, theorized by Albert Einstein. It is *bosonic (Gauge boson)* and has zero charge. Photons are the particles of *light*. Physicists have discovered the speed of light for all electromagnetic waves to be *constant* at 186,000 miles per second and this led Albert Einstein to describe the *Special Theory of Relativity*, $E=Mc^2$ where E is energy, M is mass and c is the constant of the speed of light. It interacts with Electro-Magnetism, weak nuclear force and Gravity.

Moreover, Modern Physics now explains the formation of nuclei by the process of nucleogenesis from subatomic particles, and formation of atoms, molecules, and matter. The *matter* includes both animate such as humans, animals, and insects, and inanimate such as mountains, oceans, and rivers.

Life Energy at the Most-Subtle Space

At the Most-Subtle Space, the Life-Energy remains as a *potentiality*. This Life Energy, has the potential to form many Universes continuously. Life Energy is in both animate and inanimate in the Universe.

CHAPTER 3

LIMITLESS AWARENESS

❖ ❖ ❖

AMILEA-MILELA

Amilea, Milela all-pervasive
Life-Energy inherent
In the ocean of Awareness
We are mere bubbles
In the ocean of Life-Energy
Infinite and eternal,
So, enjoy the life with
The beauty of life, with
Unconditional kindness,
Compassion, love,
Forgiveness and freedom;
No justification, whatsoever,
For frustrations of the mind,
Prejudice of all kinds,
And violence, both direct and indirect,
Isolation and loneliness.

[Amilea is the acronym for All-pervasive Mutually Inherent Life-Energy and Awareness.

Milela is the acronym for Mutually Inherent Life-Energy and Limitless Awareness.]

LIMITLESS AWARENESS

Discussion on Awareness is now presented. Even though the mind is used as an instrument in the understanding of Awareness, the limitless Awareness cannot be *contained* in the limited mind. The mind is similar to a stick that is used to kindle a camp-fire which consumes the stick; and the camp-fire remains. Sri Ramakrishna, a *Mystic* who lived in the 19th century, in India compared the mind to the stick that is used to kindle the *funeral* pyre; and the stick i.e., the mind, is consumed in the fire and the fire remains, representing the realization of *Self*, the ultimate goal of Vedanta philosophy which was also the core of Sri Ramakrishna's teachings.

The mind can be contained in the Awareness; i.e., the Awareness can be aware of the mind, as a flow of thoughts. But Awareness exists even when the mind is turned off, as in deep sleep; this is further explained below, under "The Evidence for Awareness".

Awareness and Perception

Seeing with mind:
Seeing with mind, the mind perceives everything with senses and collects the information.

Seeing with Intellect:
The intellect is capable of analyzing, investigating and inferring from the data provided by the mind; for example, all the scientific knowledge.

Seeing with Awareness:
The Awareness allows us to appreciate things beyond perception and inference. It is this Awareness with intelligence that allows us to appreciate *the Subtle and the Most- Subtle spaces.*

The Evidence for Awareness- Inherent in Life-Energy being Beyond Mind

While sleeping, the mind goes from the *'waking'* state to *'dream'* state to *'deep sleep'*. When the mind is asleep and turned off completely, the body is very much alive, with all its *involuntary* physiological functions keeping the body alive. On waking up, the mind also wakes up; and what makes it identify itself with the person, the personality and the surrounding? It is because of the *Awareness and the Life Energy* which does not go to sleep when the mind is asleep, and when the mind is totally unaware of the *'waking'* state. So how can the mind, that goes on and off, claim ownership of the Awareness that gives *life* to the mind both when it is awake and when it is asleep, and gives identity to the mind when it wakes up from the sleep state.

When we dream during sleep, the mind may be meandering in the sub-conscious state and partially functioning, similar to other systems of the body, such as heart, lungs, kidneys, and gastrointestinal system; and nevertheless, in the *deep sleep state*, even the partial functioning of the mind, commonly understood as thoughts, is absent. The mind is, in fact, a *reflection* of the Awareness. In the deep-sleep state, the essential systems, such as heart, lungs and kidneys are however, kept alive and functioning. Similarly, the brain is also kept alive, along with the memory of our identity. When we wake up, the mind wakes up, bringing the memory of our identity back to life. It is the Life-Energy inherent in the Awareness that keeps the body alive along with all the physiological systems such as cardiovascular, pulmonary, renal, neurological, endocrine and other systems, even in deep-sleep state.

Similar logic, as proof of Awareness being beyond mind, applies to anesthetized patients; and applies also to patients in coma. We have all seen or heard of people in a state of coma for a very long period and on recovery, fully regains his or her identity

In the *Theory of Mutual Inherence of Life Energy and Awareness–Milela*, I propose that the Awareness is *'Supra-mental' and is mutually inherent* in the Life Energy of all things in Nature including the animate and the inanimate; and is ever-present. The Life Energy-Awareness only *appears* to be temporarily limited by the body, during the lifetime of the person.

Quantum Mechanics and Consciousness (Awareness)

Some physicists speculate all subatomic particles may have *Consciousness (Awareness)*.

To quote physicist, E.H.Walker:

> "Consciousness may be associated with all quantum mechanical processes...since everything that occurs is ultimately the result of one or more quantum mechanical events, the universe is "inhabited" by an almost unlimited number of rather, discrete conscious, usually non-thinking entities that are responsible for the detailed working of the universe"

Time-Line and Proof for the Life-Energy Inherent in Awareness Being Beyond Mind

A review of *time-line*, based on archeological and fossil studies, shows that the origin of *Homo Sapiens*, the ancestors of the present-day human beings, dates back to only 300,000 years. The common ancestor of *sapiens* and Neanderthals (*Homo neanderthalensis*) is known as *Homo heidelbergensis*. Study of the fossils and their genomes shows that *Sapiens and Neanderthals* split from their common ancestor approximately 500,000 years ago. Sapiens attained their full-fledged form only 300,000 years ago, and Neanderthals attained their full-fledged form, 250,000 years ago.

Plants and animals came into existence, about 200 million to 300 million years ago, millions of years before the *Homo sapiens*. *Synapsids* resembling the present-day reptiles and birds appeared 320 million years ago. *Jurassic period*, made famous by the movie of a similar name, is the second segment of *Mesozoic era* and dates back to 199.6 million years to 145.5 million years ago. Reptiles, *Dinosaurs* dominated the Jurassic period. By the beginning of this period, plant life had evolved from *Bryophytes*, the low growing mosses.

Brain development started with only the sense of smell, about 200 million years ago. The existence of *Life Energy Inherent in Awareness*, of course had to be present before such *brain development and its individualized mind*. For example, in comparison, the Universe existed billions of years before the life forms with any trace of mind, discussed above. According to the Milela theory, Life Energy-Awareness pre-dates (if there was time and date at that time!), even the *Big Bang and the origin of Universe*. Life-Energy inherent in Awareness is the basis of the origin of Universe. *Big-Bang* is dated to be 13.7 billion years ago. Cosmologists estimate the origin of the Universe to be 4.7 billion years ago; and the *Dark Energy* started to accelerate 3.7 billion years ago. So, Life-Energy inherent in Awareness predates the origin of mind as shown above, in the time-line; and so, is definitely *beyond mind.*

Proof of Life Energy in the Plant Kingdom

Plants having Life Energy is further established by comparing the plant cell with human cell and an animal cell.

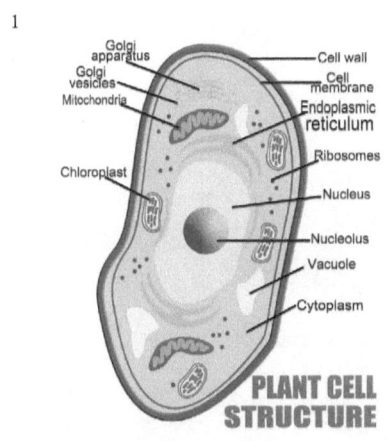

1. Samuel, Leslie. *Plant Cell Structure.* Accessed May 25, 2018. http://www.interactive-biology.com/3956/anatomy-of-the-plant-cell-vs-a-human-cell/

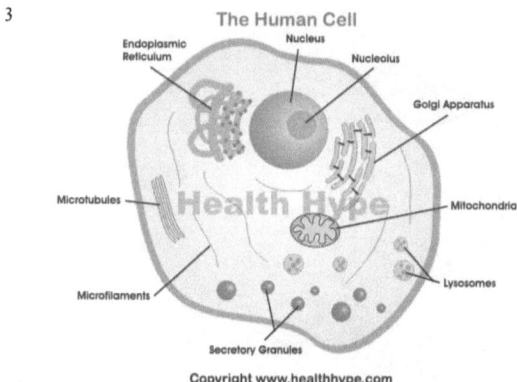

2 *Animal Cell Anatomy.* Accessed May 25, 2018. http://www.enchantedlearning.com/subjects/animals/cell

3 *The Human Cell.* Accessed May 25, 2018. https://www.health-hype.com/human-cell-diagram-parts-pictures-structure-and-functions.html

The cell structure and function of plants have an amazing similarity to the cell structure and function of human beings and mammals!

Except for the photosynthesis and the exchange of O_2 and CO_2 being opposite, the similarity is remarkable. The cell is the basic unit of all life that is *eukaryotes* with nucleated cell. As shown in the diagrams, all these cells contain nucleus, nucleolus, cytoplasm and cell membrane. Within the cytoplasm, the components include mitochondria, lysosome, ribosome, and ATP as the chemical to carry energy.

In the functions of the plant cell also, such similarity continues on. The DNA (*Deoxyribonucleic acid*) within the nucleus contains the genetic code and produces RNA (*Ribonucleic acid*) which carries the genetic information to the *ribosomes* in the cytoplasm for production of proteins and enzymes. Ribosomes, estimated to be about 500,000 in each cell, contain RNA.

Lysosome is described as the recycling-center with enzymes to digest food and other cell components and to recycle to form new cell components. Mitochondria are the *battery* for the cell. They produce energy from the broken-down sugars and oxygen from *photosynthesis* that takes place in the *chloroplast*.

Chloroplasts take in CO_2, H_2O and solar energy and produce glucose and O_2. Oxygen is given out and the sugars are taken by mitochondria to produce energy. This energy is carried by ATP, (*Adenosine Triphosphate*) to other parts of the cell for supplying energy.

The above example of plant life, human and animal life, and environment interacting with each other with exchange of CO_2 and O_2 in opposite directions proves that Life Energy is Universal. Moreover, such complex functions and interactions of Life Energy can only happen with the presence of Universal intelligence or Awareness.

FURTHER PROOF FOR 'AWARENESS INHERENT IN LIFE-ENERGY' IN THE PLANT KINGDOM

Plants do not have an anatomical brain and eyes to see nor any hearing apparatus.

Yet we see plants growing up or to the side towards any direction where there is sunlight. If there is a building structure close to the plant it bends away from the darkness towards the sunlight. It may also straighten up after reaching the area of sunlight. Sometimes the plants may even grow between the rocks arising from some soil underneath the rock or from the rock itself, reaching for the sunlight. How does the plant know the location of the available sunlight without brain or eyes? It is from the Awareness inherent in the Life Energy itself of the plant. In other words, the Life Energy of the plant is Inherent in the Awareness.

After I recognized the Awareness being *beyond mind* with proof given under the *Milela Theory*, I reviewed the recent literature on this subject. Gregory Bateson came very close, by the recognition of cognition being beyond mind by his conclusion that *"cognition is a mental process outside the brain"* and mind exists both in the brain and outside the brain.

Maturana and Varela described self-organizing and self-making of *living systems* as *Autopoiesis* in which living systems function as a network pattern in which each component takes part in the self-making of other components of living systems. Maturana and Varela also expanded the cognition as part of the living system. However, both studies confined the *Autopoiesis* and *cognition as a mental process*, as characteristic of living systems only. They did not include the *inanimate*. Besides, cognition is described by Bateson as a mental process, even if they function outside the brain.

Fritjof Capra reviewed and concluded that the studies above that came to light in the 1960s further helped to overcome the Cartesian division between mind and matter; 'the interactions of all living organisms with its environment are cognitive or mental interactions, inseparably connected to *Life*.'

However, both life and cognition as a mental process were confined to living systems only, in the studies and review mentioned above. This limitation to the living systems only continued on until *Gaia*

Theory came along. In Gaia Theory, James Lovelock, a chemist and Lynn Margulis, a microbiologist, proposed that the whole *Earth* is a *living thing* with its components interacting with each other. The quantum theory came along in 1900 CE, and further developed in the twentieth century, to show that energy in momentum, appearing as the subatomic particles, is the basic unit of life and all things in Nature; this was discussed in detail earlier, under Life Energy and life.

Awareness and Cognition

Cognition is sometimes confused with Awareness. Cognition is not Awareness, even though, Awareness is the basis of all functions of mind and body, including cognition. Generally speaking, any structure or function that has the quality of changing or is ephemeral cannot be the *limitless Awareness*. Cognition depends on Awareness for its existence. Cognition is a part of the mind; and cognition aspect of the mind is connected to the brain at the synapses. For example, in *Alzheimer's disease*, there is impairment of cognition, either mild or severe, with loss of memory.

In the studies of patients with Alzheimer's disease, *synapse-loss* was shown. Robert D. Terry, MD et al published an original article in "Annals of Neurology" in 1991, a prospective series of 15 patients with Alzheimer's disease and 9 controls. It showed loss of synapses in the brain, measured by 'immunocyto-chemical- densitometric technique'.

CONCLUSION

In conclusion, I have proposed the *Milela Theory* which stands for *Mutual Inherence of Life Energy and Limitless Awareness* with adequate proof, as the *general theory* of all things in Nature; and *Chetana Yoga or Chaitanya Yoga* as the method of practice of Limitless Awareness, as the next step in human evolution.

UNIFICATION THEORIES

In Modern Physics, many theories have been proposed to *unify* all the *Fundamental Energy Forces of Nature*. Most notably, there are four:

1. GUTS–Grand unified theories, unifying three of the four Energy Forces of nature, namely *Electro-Magnetism Force, Strong Nuclear Force and Weak Nuclear Force*. Gravity is the other Energy Force.
2. SUPER-SYMMETRY- This theory proposes the presence of *sparticles*. It is theorized that the *sparticles* interact and cancel massive amounts of particles, such as photons and

electrons that constantly appear 'in and out of existence'. However, presence of *sparticles* has not been proven in any of the particle physics-laboratories.
3. MULTI-DIMENSIONAL THEORIES- As the name implies, it proposes multiple dimensions with *Gravity* flowing through them and getting weaker and weaker in that process. This explains the Gravity being weak in our world in comparison to the *Origin of Gravity*.
4. FLIPPED Su(5)-combines GUTS and Super-Symmetry.

All of the above four theories have not been proven in any of the particle physics laboratories such as CERN (European Organization for Nuclear Research), which has the LHC-Large Hadron Collider- and is building more LHCs, near Geneva, Switzerland. However, the fact that they are not proven in the laboratories, does not rule out their validity. They are all probably true, explaining the different aspects of the reality of the Energy Forces. If we relied only on laboratory confirmation, science and the world would have missed out, on many discoveries. Isaac Newton's mind would not have noticed the existence of Gravity, on seeing the apple falling down from the tree. He then pursued the theory of Gravity, based on his *observation*. James Watt would not have *observed* the power of steam, when his mother was boiling the water in a tea-kettle. The steam engine would not be a reality today.

Hans Selye (1907-1982 CE) as a young physician *observed* the common thread of *stress* in all the patients with different diseases and came up with *Stress Syndrome or General Adaptation Syndrome*. Further research led to the finding of increase in *cortisol* from adrenal cortex in *stress*.

To quote Hans Selye, stress syndrome "could have been discovered during the middle-ages" –and its "recognition did not depend upon the development of any complicated pieces of apparatus, but merely upon an *unbiased state of mind*, a fresh point of view".

Milela–as the General Theory of all Things–the Unifying Theory

All four of the *unification theories of Energy Forces*, mentioned above, have one thing, in common, namely they are trying to explain the Energy forces from the level of the particles upwards to the level of Gravity and Electro-magnetism. However, there is another way of approach from the level of the Gravity and Electro-magnetism downwards to the *subatomic particles*, as summarized below.

The Life Energy-Awareness Inherence is now proposed as the *Unifying Energy of All Things* in the Universe including the Fundamental Energy Forces of Nature. In Physics, the Energy Forces of Nature are known to consist of:

1. Gravity
2. Electro-Magnetism
3. Strong Nuclear Force
4. Weak Nuclear Force

According to Milela Theory, Life Energy inherent in Awareness-Space *expands* to become *Gravity-space*, as a process of circumferential expansion; and so, the Gravity-space is spherical to begin with and the Electro-Magnetism is linear. The spherical Gravity becomes oval-spherical when it interacts with the linear Electromagnetic energy. The spherical character of the Gravity Electro-Magnetism *duality* continues in all its manifestations, such as *subatomic particles, atoms, cells, planets, stars, galaxies, and even space-time all of which show the same spherical or modified spherical character.*

Gravity and Electro-Magnetism duality coexist continuously in the Subtle Space as well as in the Least Subtle Space similar to the Mutually Inherent Life Energy and Awareness. However, according to *Milela, and the Theory of Continuous Subtlety* (to be described in the next chapter), where Gravity is predominant, Electro-Magnetism is less predominant and where Electro-Magnetism is predominant, Gravity is less predominant.

Gravity and Electro-Magnetism, born out of MILELA (Mutual Inherence of Life-Energy and Limitless Awareness) continue their duality *coexisting* all the time down to the subatomic particles that form the basic unit of all things in Nature. The subatomic *particles* are also known to behave as *'waves'*. Further, it is this author's opinion that it is Gravity that gives the *particle aspect* and it is Electro-Magnetism that gives the *wave aspect* of the subatomic particles. As a result, it is Gravity that gives the particles, the *position*; and it is Electro-Magnetism that gives the *momentum* of the wave. Moreover, considered separately, Gravity-Space is the origin of *Space;* and Electro-Magnetism is the origin of *Time*. *The coexistence of Gravity and Electro-Magnetism* is the reason for the coexistence of *Space-Time continuum.*

It is a known fact in physics that Gravity does not have a *'particle'* associated with it; whereas everything from Electro-Magnetism onwards is proven to be associated with a particle.

Milela theory is presented as the *General Theory of all things*, because it satisfies the need for such a theory, to be *complete* and *correct* in the microscopic world of Quantum Physics as well as in the macroscopic world, we live in. Moreover, Milela as the General Theory of all things is explained as a reconciliation of different branches of science and philosophy. All the *divisions* of knowledge are thus made by us, human beings, for the sake of understanding and specialization. For example, our body does not say, 'I will do my bodily functions of organic chemistry from 8 to 9am, of physics from 9 to 10 am, of microbiology from 10 to11am, endocrine system from 11 to 12 noon–lunchbreak–and then neurology from 1 pm to 2 pm etc. The body functions as a *whole;* so, it is only reasonable to demand a theory of the *whole*, including this world and the whole Universe.

MILELA AND EINSTEIN'S GENERAL THEORY OF RELATIVITY

As mentioned above in the Milela theory, the spherical or global character of Gravity Electro-Magnetism duality is carried on to space-time. It is this spherical character of the space-time that gives the *warping of space-time* presented by Albert Einstein in his General Theory of Relativity. In other words, the *warping of space-time continuum* is an *arc* in the oval-spherical character of the Gravity Electro-Magnetism duality and the space-time continuum.

The spherical character of Gravity Electro-Magnetism duality continues on from the Subtle Space that is considered by Physics to be 96% of the Universe, to the Space we live in, that is 4% of the Universe known to Physics of today.

MILELA AND THE MILKY WAY GALAXY

Milela speculates that the duality of Gravity Electro-Magnetism continues on to the Dark Matter and the Dark Energy of the Subtle Space. Gravity is more predominant in the Dark Matter and the Electro-Magnetism is more predominant in the Dark Energy. As a result, in Dark Matter, Gravity is more predominant and Electro-Magnetism is less predominant, maintaining the *law of proportionality*. For the same reason, in Dark Energy, Electro-Magnetism is more predominant and Gravity is less predominant.

Milela further speculates that the Milky Way galaxy is in the periphery of the Dark Energy where Gravity is of minimal predominance, explaining the *relatively weak force of Gravity* in our world. This explains why the Gravity becomes weaker and weaker and the Electro-Magnetic energy becomes stronger and stronger, as we travel up into the space.

This theory also proposes that the All-pervasive Life Energy and Limitless Awareness, although named and described separately, are inseparable, mutually inherent, non-dual and one and the same—hence it is named as the *Mutually Inherent Life Energy-Limitless Awareness–THE MILELA*.

Section Five:

ORIGIN

CHAPTER 1

SPACE AND SUBTLETY

❖ ❖ ❖

Theory of CONTINUOUS and RELATIVE SUBTLETY

A tree says: A kernel is hidden in me, a spark, a thought, I am life from eternal life. The attempt and the risk that the eternal mother took with me is unique, unique the form and veins of my skin, unique the smallest play of leaves in my branches and the smallest scar on my bark. I was made to form and reveal the eternal in my smallest special detail.

— HERMANN HESSE 1877-1962

GRAVITY IS SPACE. THE *AWARENESS-SPACE expands* to become the Gravity space. The Awareness Space *(known as Chith-Ākasha* in Sanskrit) with its Life-Energy potentiality *desires* to expand (the *desire* being the first of the triple-principle of *Milela)*. The Awareness provides the *knowledge* (the second of the triple-principle), leading to the *action* or manifestation of Gravity-space and Electro-Magnetism *(action* being the third of the triple-principle of *Milela)*, described earlier.

The Awareness-space with its Life-Energy potential only *appears* to expand! Nevertheless, it always remains as *Awareness-space*. This *expansion* is circumferential, all around, from the *Life-Energy potential;* so, the Gravity-space appears to be spherical. It is of interest to note that there is *no known particle for Gravity*, even though *Gravitron* is suggested as the particle of gravity.

The Life Energy of Milela, as the *Most-Subtle Awareness space*, giving rise to the duality of Gravity and Electro-Magnetism, at the *Subtle space* level, continues to be the Gravity-space; and Electro-Magnetism with its tremendous movement contributes to the concept of time, thus leading to the *space-time continuum*. According to the *Milela theory*, the interaction of the *attractive force* of the Gravity-space along with the *attractive and repulsive forces* of Electro-Magnetism, forming opposite forces at a tremendous speed, produce trillions and trillions of *subatomic particles*. It is well known in Physics, that the subatomic particles include negatively charged electron which is a *particle* and positively charged positron which is an *anti-particle* leading to early annihilation. Other particles include positively charged proton and neutron (contributing to form the nucleus of the atom,) and recently discovered bosons. The trillions of electrons and positrons undergo *early annihilation* producing two *photons shooting out* at the speed of light, for each electron and positron.

The interactions, described above, lead to the formation of *atoms* with the electrons orbiting around the nucleon, formed by proton and neutron. The atoms again are *mostly space* with a huge span of space between the orbiting electron and the nucleus with its positively charged proton and the neutrally charged neutron.

The particles themselves are in momentum constantly and *appearing* in perception of the observer only, due to the tremendous movement, nearly the speed of the light (the speed of light constant, being 186,000 miles per second), also appearing and disappearing, similar to a magic show *now you see, now you don't!* A possible explanation of this phenomenon of *appearance and disappearance of the subatomic principles* is given as follows:

If the momentum of the particles exceeds the speed of light (*super-luminal*), they disappear in the *Awareness-Space*, only to reappear again at the momentum of speed of light or lower! The atoms are the basic unit of all things in the Universe. We, human beings are no exception to this reality; and so, we are also mostly *space*. The subatomic particles accelerating at a tremendous speed nearly the speed of light, beyond our perception, thus *appear as forms*.

It is interesting to note the similarity and continuity in the Sanskrit language, of words for Awareness, Gravity and Space, suggesting that Gravity-space is derived from Awareness-Space (with Inherent Life Energy); and Gravity is the space in the space-time continuum. The word for Awareness in Sanskrit is often referred as *Chith-Ākasha*, *chith* meaning Awareness and *ākasha* is Space. *Ākarshana* means *attracting*, and *Ākarshika* means *attractive* as well as *magnetic*. The word for Gravity is *Ākrishti-Sakti*.

SUBTLETY

Subtleness or Subtlety is defined in the Merriam-Webster Dictionary as "difficult to understand or perceive— *obscure*." Other meanings given include "hard to notice or see: not obvious."

In Sanskrit language, the Subtlety is called *Sukshmam*. In Tamil, *one of the two classical languages* of India, it is called *Maraithal*, meaning it is obscure.

From the Mysticism point of view and this author's point of view, *subtlety* means *intangible*. Even the 'tangible' things of the Universe are in fact intangible. From the view point of atoms and subatomic particles which are the basic unit of *all things in Nature* also, the tangible things are intangible. They are so small that it needs the most sophisticated instruments. Because the subatomic particles behave both as particles and waves, they have both a *position and movement;* however, if the observer decides to observe the *position* fully, the movement is not seen; and if the observer decides to observe the *movement* fully, the position is not recognized. So, it depends on the observer's choice whether to perceive the position or movement. It is not possible for the observer to see both the position and the movement, same time.

The spiritualists and philosophers have been calling the subtlety as *illusion*. From the view point of the theoretical physicists, according to the Quantum Theory, the subatomic particles are present in perception only.

Summarizing the different viewpoints of the subtlety, all things in the Universe *appear* to have a form. Where does the appearance of the forms come from? It is due to the rapid *movement* of the subatomic particles nearly the speed of light that is characteristic of Electromagnetism. So, all things in Nature, as the subatomic particles, are in *continuous movement* beyond our perception because it is at a tremendous speed, nearly the speed of light. It is similar to a toy-fan which children run with, making the blades of the toy-fan swirl at a rapid speed, giving the *appearance* of a circle or globe. So, on summarizing the above discussion, the *Space along with its subtlety seems to be the only reality*.

The Dancing Siva

The Eternal Dance
What a gift of beauty, Nature bestowed
With love, kindness and free-will
Inherent in Life-force, inner nature
What a waste not to hold the heavens
In the chambers of the heart
Subtle and eternal, yet within reach
As Awareness
Good intentions
True conscience
Kindness, love, and freedom.
Super-luminal dance, subtle and mystical
With no beginning no end
The eternal dance.

The *Dynamic Cosmic Dance* is described as the *Dancing Siva or Nataraja* in the Mythology of India. I must add that according to the *theology of India*, the Dancing Siva, representing the potentiality of Life Energy is one of the *aspects or facets* of the formless Supreme Being known as *SIVAM*.

Origin

The association of Dancing Siva to our present-day knowledge of physics and subatomic particles have also been eloquently elaborated by Fritjof Capra in his popular book titled "The Tao of Physics." The Dancing Siva represents different things to different people. It is the representative form of the *formless Sivam*, the ultimate supreme being for many devotees. For many devotees and spiritualists, the dance is the *cosmic dance*. The surrounding *circle*, known as *Thiruachi* in Tamil, represents the Consciousness or Awareness; and the form inside, the dancing pose, represents the power of Siva with five elements, *space, air, fire, water and earth elements* constituting the whole Universe. The little dwarf under the right foot of the Dancing Siva represents ignorance. The right hand is showing the *abhaya mudra*, saying *'fear not'*. The left-hand crosses over to the right side, just below the right hand and denotes *protection* to all. The left leg raised up and crossing over to the right side, in the dancing pose represents the *balance* maintained in the Universe.

To reconcile the present-day knowledge of Modern Physics and the Milela theory with the Mythology of India, the continuous Dancing of Siva, representing the *Potential Life Energy and Awareness Space*, leads to *Gravity-Space and Magnetic Energy*, known as *Ākarshana Shakti and Kānta Shakti* respectively. Kānti, a word closely related to Kānta, in Sanskrit means *brilliance*, indicating that the Magnetic force also represents Electro-Magnetism.

Moreover, Ākarshika in Sanskrit means both attractive force and magnetic force, leading to the concept of the co-existence of Gravity and Electro-Magnetism.

Magnetic Energy in Physics was later expanded to Electro-Magnetic Energy by Maxwell's theory of Electro-magnetic energy.

The tremendous speed of the dance is *superluminal*, meaning it is much greater than the speed of light beyond our comprehension. As discussed earlier, Gravity is *space* and gives rise to *space; and Electro-Magnetism* and its continuing movement gives rise to *time*. Only when the tremendous speed of the movement slows down to the speed of light or lower, with the interaction of Gravity and Electro-Magnetism, the Energy turns into appearance of the subatomic particles, such as electrons, protons and neutrons (forming the atoms) and positrons, and photons (the light particle), in the *space* leading to formation of atoms, molecules and forms.

The dance continues on, as the dance of the subatomic particles. The Most-Subtle Life Energy-Awareness becomes the Subtle; the limitless becomes the limited; the formless appears as the forms. Energy becomes mass, as proven by *Einstein's Special Theory of Relativity* $E=Mc^2$ (c is the speed of light *constant*).

Einstein's theory also said that kinetic energy has mass; so, energy is mass and mass is energy. Energy and mass are different forms of the same thing.

Energy is light energy when the movement of the Electro-Magnetic force is at the speed of light; and the other particles appear at lower than the speed of light (which is constant at 186,000 miles per second). The number of protons in the nucleus determines the type of mass. In other words, energy–the Life Energy of *MILELA* at the Subtle as well as at the Least Subtle (gross) space levels–is *subtler* than mass.

THEORY OF SUBTLETY AND PHYSICS

At the present-day knowledge of physics, it is a well-known scientific fact that the subatomic particles are always moving at a tremendous speed and the speed varies for different particles. It is also this movement that gives them the *wave-like* quality to the particles.

According to quantum physics, the speed of each subatomic particle is constant; for example, the speed of all protons is the same. It is the number of protons that determines the form. Nevertheless, to reconcile with the *Theory of Subtlety*, it is further suggested that at the *Least Subtle level*, depending on the speed of the movement of the subatomic particles, at varying speeds for different particles, they become less and less *subtle*, reaching the *Least Subtle State* of existence where the different *forms* become apparent within the Universe including atoms, molecules, air, fire, water and earth, galaxies, stars, and planets. The number of protons in the nucleus determines the different forms. On the other hand, at the *Subtle level*, when the speed of the movement of the subatomic particles increases, they become more and more *subtle*, reaching the subtle state of existence, losing the appearance of forms and becoming pure energy of *Electro-Magnetism* and *Gravity*. When the speed of the subatomic particles further exceeds the speed of light constant (186,000 miles per second) –*super-luminal*–it remains as the *potential Life-Energy*, inherent in the *Limitless Awareness*, herein described as the *Most-Subtle Space*.

Physicists have theorized *super-luminal light force* (faster than the speed of light). However, the exact nature or identity of *super-luminal phenomenon* is yet unknown to the Physics-world.

The Theory of Subtlety

Similar to the above discussion of *space*, understanding the three *states of Existence* is essential. The three states are:

1. the Least Subtle or the gross State
2. the Subtle State and
3. the Most Subtle State

The theory of continuous and relative subtlety is proposed as follows:

There are continuous and relative subtlety states of *existence* from *the Least-Subtle to the Subtle and the Most-Subtle*.

The less and less the *subtleness*, there is more and more delineation to establish a *form, within the space and time continuum*. The more and more subtlety of objects in the universe, there is less and less limitation continuously, until only the all-pervasive *Most-Subtle Limitlessness* prevails. For example, the liquids are subtler than solids. Air and gases are subtler than liquids. Air and gases exist in the space; and the space is subtler than air and gases.

Space and Subtlety

The theory is further elucidated as follows:

The Space is not 'ether' which was a concept, now abandoned by Physics. The *Space* consists of the

1. Least Subtle or gross space
2. Subtle space and
3. Most Subtle space.

The *Least-Subtle space* perceived by us, as in the world as well as in the outer space in the universe, is contained in the *Subtle Space*; and the Subtle space is

Subtler than the Least-Subtle. The Subtle Space of Gravity and Electro-Magnetism originate from the Most-Subtle Space, as described earlier.

No Border

There is no border or demarcation between the Subtle space and the Least-Subtle space (or gross space), we live in, because space is *continuous and relative.*

There is only *space* where all the *dance* of the various subatomic particles takes place with creation, annihilation and rebirth of particles continuously with the interaction of Electro-magnetic force and Gravity. Quantum mechanics prevails in both the Subtle space and the Least-Subtle space and is the manifestation of the *reality* but cannot give an understanding of the *reality* itself. Nevertheless, the quantum mechanics is not visible in our world because the subatomic particles are very, very small and they are perceived to be in *perception only*, according to Quantum Mechanics.

In the Copenhagen Interpretation of Quantum Mechanics that was formulated in 1927, the Physicists recognized that a complete understanding of *reality* lies beyond the reach of rational thought, in other words, beyond the logical mind. This was further clarified by the quotes of advanced physicists, shown in the popular book, "The Dancing Wu Li Masters" by Gary Zukav as follows:

Origin

Henry Pierce Stapp, theoretical physicist, Lawrence Berkeley Laboratory in Berkeley, California, in his own words:

The Copenhagen Interpretation of Quantum Mechanics "was essentially a rejection of the presumption that nature could be understood in terms of elementary space-time realities. According to the new view, the complete description of nature at the atomic level was given by probability functions, that referred, not to underlying microscopic space-time realities, but rather to the macroscopic objects of sense experience. The theoretical structure did not extend down and anchor itself on fundamental microscopic space-time realities. Instead it turned back and anchored itself in the concrete sense realities that form the basis of social life. ...This pragmatic description is to be contrasted with descriptions that attempt to peer behind the scenes and tell us what is really happening."

Professor David Bohm, Professor of physics, Birkbeck college, University of London in a lecture given in University of Berkeley in April 1977, proposed,

> "we must turn Physics around. Instead of starting with parts and showing how they work together we start with the whole."

In a lecture given in the Lawrence Berkeley Laboratory to the Physicists, also in April 1977 Professor Bohm touched on the *subtle mechanism* needed: "The ultimate perception does not originate in the brain or any material structure, although a material structure is necessary to manifest it. The subtle mechanism of knowing the truth does not originate in the brain."

Further, Gary Zukav has also quoted Professor G.F. Chew, Chairman at Physics department, University of Berkeley, making the following comment in reference to a theory of particle physics:

> "Our current struggle (with certain aspects of advanced physics) may thus be only a foretaste of a completely new form of human intellectual endeavor, one that will not only lie outside physics but will not even be describable as "scientific."

Our world, however, for all practical purposes, still follows the laws of Classical Physics a.k.a. Newtonian Physics.

Life Energy potential, inherent in Awareness, being the Most-Subtle Space as the *reality*, pervades all space including the Subtle space and the Least-Subtle space.

THE MOST SUBTLE SPACE
The Least-Subtle and the Subtle space are both contained in the *Most-Subtle Space;* and so, the Most-Subtle

Space being *all-pervasive* is subtler than the Subtle space and the Least Subtle space.

According to the Modern Physics, the Least-Subtle or gross space and the Universe that we know of constitute only 4% of the Universe; and what is unknown is the remaining 96%. This 96% of the Universe that is not known to the present-day knowledge in physics, is actually the *Subtle Space*.

With reference to the theory of *Mutually Inherent Energy and Awareness* that is described under the *MILELA Theory* in the previous chapter, Energy and Awareness are both inherent and inseparable from each other. Such *Energy and Awareness* constitute *the Most-Subtle Space* that is *all-pervasive* existing in all the states of existence and at the same time, are the basis or substratum of *All Space*, including the Least-Subtle and the Subtle spaces. This *subtlety* is present in everything and everywhere all the time, from the Least Subtle to the Most-Subtle in which the Life Energy remains as the *potential force*.

The Subtle Space

The Subtle space is the space that is unknown or not clearly understood in our present-day knowledge and constitute 96% of the Universe. What is known is 4% of the Universe.

The Subtle space is the most fascinating of the three spaces, spewing out trillions of subatomic particles like a *fourth of July fireworks show*, multiplied zillion times. It is similar to one of the fireworks that goes up in the space and forms a globe of multiple star-like lights and more globes of light spring up from the first globe!

The saying, 'Truth is stranger than fiction' aptly applies to the magic show of the *Subtle space*, even though it is only a fraction of the manifestation of the *Reality*.

The magic of the *dance* of the Subtle space continues on and on with formation, annihilation and rebirth of subatomic particles. Moreover, the mystery of *Black-hole* is being unraveled to some extent in Physics. The Black-holes are considered to be burned out stars, packing in millions of times the energy of sun within a smaller space than the sun. So, it has the most-dense gravitational force, swallowing all that comes within its vicinity and not allowing even a trace of light to go out. All galaxies supposedly contain a black-hole in the middle.

The fate of the planets or galaxies swallowed by the black-hole is still a mystery, speculated to come out on the other side of the black-hole through *Quasars*–quasi-stellar radio source–sometimes referred to as the white-hole, into a different Universe! Alternately, the galaxies may come out of the black-hole itself, through *worm-holes* in a different *space-time* into the same Universe! The galaxies may also enter through a worm-hole and come out through a worm-hole.

Perhaps, there is a teleological purpose or need for the galaxies to get *re-energized*–recharged so to speak–in the black-hole, this way!

THE LEAST SUBTLE SPACE

The same *Inherent -Awareness-Energy*, at the Least-Subtle (or gross) level exists *as the Life Energy*, in all things in Nature.

This Life-Energy, is present everywhere in both animate and inanimate of the whole *Universe*.

It exists as Life Energy or Life Force, at the *individual level*, keeping us all alive, in the same way as Life Energy for the *whole Universe*.

This Life Energy, *as the Unifying Energy Force* contains the Fundamental Energy Forces of Nature, at the *Least Subtle Space* as well as at the *Subtle Space* including

1. The Force of Gravity
2. The Electromagnetism
3. The Strong Nuclear Force and
4. The Weak Nuclear Force.

The same Inherent-Awareness-Energy exists at the Least-Subtle (Gross) space as the body and mind along with the Life energy at the *Individual level*. However, *the mind is more subtle, relative to the body.*

And the same Inherent-Awareness-Energy exists as all the matter, including solids, liquids and gases and earth, water, fire and air and energy forces, such as solar energy, electrical energy and geothermal energy, and exists as radioactive cosmic rays at the *Universal level*.

This Theory of Relative and Continuous Subtlety can be explained with a simple example with reference to water. The ice-block is the solid matter; when it melts, it becomes the liquid water. So, the liquid water is subtler than the ice block. When the water evaporates to hydrogen and oxygen, they are a part of air which is subtler than the liquid water. The air blends with the space which is subtler than the air.

In the modern physics, after Einstein presented the 'special theory of relativity',

$$E=Mc^2,$$

where c is the speed of light constant, we now understand that energy can be converted to mass and mass can be converted to energy at the speed of light squared; the conversion does not actually happen but it proves that *energy and mass are different forms of the same thing*. Even in the solid matter, itself, there is a relative subtlety present, as shown in the example of ice block and water.

Applying this *Theory of relative and continuous subtlety*, the molecules of a thing or matter is subtler than the thing itself. The atoms that form the molecules are subtler than the molecules. The subatomic particles are subtler than the atom.

All subatomic particles are in fact energy in different forms. Between the subatomic particles, electrons are considered subtler than protons. The electrons are known to be present everywhere. Similarly, the *quarks* within the protons are subtler than the protons.

Recently discovered Bosons, known as the Higgs-Boson Particles, now understood to be the most basic energy particle responsible for all matter, are subtler than the electrons. This relative subtlety is present all the way from the Least-Subtle to the Most-Subtle in everything and everywhere all the time.

The subatomic energy particles such as electrons and bosons which form the basic units of all things in Nature, and which, according to quantum theory, exist in perception or *Awareness* only, constantly appearing and disappearing, are providing a living proof of the *Theory of Continuous and Relative Subtlety.*

At the *subtler* level, it is understood that *energy*, as the subatomic energy particles, exist in *Awareness*, even though they are not visible to perception at the solid level in everyday life, showing that *Energy and Awareness are Inseparable and thus Inherent* of each other.

When this understanding is carried back from the subatomic energy particles, such as electrons to solid matter also, we realize that the *Energy* and *Awareness* exist continuously from subtle to solid matter; and the Life energy and Awareness exist continuously, *inseparable* of each other.

In fact, *subtlety* is the only reality. The Milela-Theory described in the previous chapter says that the Mutually Inherent Life Energy with Limitless Awareness is the *reality behind all existence* and it is the Most-Subtle. Next comes the Subtle, which includes Dark Matter and Dark Energy, which together are known to constitute 96% of the Universe according to the physicists. The remaining 4% is the Universe we know of, and it is the Least-Subtle or the gross space which is within our *perception* and/or knowledge.

Another example to prove the presence of subtlety as the reality is the three states of existence, namely *wakeful state*, as the Least Subtle state of existence; and *the dream sleep*, as the Subtle state of existence; and *the deep-sleep* where the person's mind and the wakeful state are not present, as the Most-Subtle state of existence.

Between the body and mind, the body exists at the solid level and the mind exists at the subtle level, relative to the body. Both the body and mind, however, belong to the Least Subtle or Gross space, as distinguished from the Subtle and the Most-Subtle spaces. There is also connection between the body and mind, explained in the chapter on Body-Mind Connection.

CHAPTER 2

ORIGIN OF ALL THINGS IN NATURE

> "Give me a place to stand,
> and a lever long enough, and
> I will move the world"
>
> — ARCHIMEDES

ORIGIN OF ALL THINGS IN nature have been an enigma and has been described in many different ways. Trying to find the origin of all things is like trying to find the ultimate inside of the onion bulb; the more you peel the peels, you end up with *nothing*. The *nothing* as well as all that is related to the *origin* are only in perception or Awareness.

Poets and philosophers have been calling it *illusion*, *super-imposition* and *Maya*. Spiritualists also have been saying that it is *Maya* and is inscrutable and indescribable. Religious leaders depend on faith in the belief system of a creator.

In the Modern Physics a.k.a. Quantum Physics, the theoretical physicists have proven the existence of subatomic particles as the basic unit of all things in nature; and they seem to behave as both *'quanta' or packets* of *particles and waves*. Einstein's Special Theory of Relativity, $E=Mc^2$ again proves that all mass or matter is another form of energy in *movement* where c^2 is the constant speed of light squared. Einstein had shown that the kinetic energy has mass and mass is energy in movement. *Energy and mass are different forms of the same thing.*

Now, the *Milela Theory* of Mutually Inherent Life Energy in Limitless Awareness is described here with adequate proof as the *General Theory of all things* and the unifying theory to bring it all together, bringing philosophy, spirituality, religion, and science towards *one ultimate reality* as well as to unify the fundamental forces of energy.

The purpose of this book is mainly to establish the Milela-Theory as *the Most-Subtle* and as the *All-Pervasive Energy inherent in Awareness*, existing also in the *Subtle* and *the Least Subtle States* (also known as the Gross State) of the Universe that includes all things in the Universe including this world in the milky way galaxy, and billions and billions of galaxies. It is also proposed that the Life-Force inherent in the Limitless Awareness plays a role in the Origin of all things in the Universe.

This Universal Awareness-Energy is indeed the *Origin of all things in Nature and* the *Origin of the fundamental forces of energy* which is understood in Physics to consist of,

1. Gravity
2. Electro Magnetism
3. Strong Nuclear Force
4. Weak Nuclear Force

From Physics point of view as well as Philosophy point of view and in particular this presentation of 'Origin of all things in Nature' point of view, the main fundamental forces of Energy arising out of the Universal Awareness-Energy are *two*.

1. Gravitational Space (also known as Gravity Interaction) — *Ākarshita or Ākarshana Shakti* in Sanskrit, means *Gravity Force that attracts*; and it is interesting to note that another related word in Sanskrit, *Ākarshika* means both *attractive force and magnetic force*.
2. Magnetism (also known as Electro-Magnetism)— *Kānta Shakti* in Sanskrit means *Magnetism or Magnetic Energy*. Another related word, *Kānti* in Sanskrit means Brilliance.

In 1864, Maxwell's equation and the theory of Electromagnetic field were described and the term, Electro Magnetism was adopted to describe the Magnetic Force.

Gravity is known to possess attractive force. Electro-Magnetism, on the other hand, has both attraction and repulsion forces. The opposite charged (negative and positive charged) particles attract; and the similar charged (both negative or both positive charged) particles repel.

From *Mysticism* point of view, the two opposing forces of attraction and repulsion begin the churning action at tremendous speed, resulting in the production of trillions of subatomic particles.

The trillions and trillions of electrons and positrons undergo an *"early annihilation"*. The electrons prevail in the early annihilation, in destroying the positrons, with the help of Electro-Magnetism. Physicists speculate that slightly more number of electrons were produced, helping the electrons to prevail. When electrons (particles) and positrons (anti-particles) come together, both are destroyed and replaced by two *photons* which shoot out at the speed of light. The subatomic particles go on to form the trillions of atoms and billions of *galaxies and stars*. They are pushed away from *Dark Energy* and eventually to the *Least-Subtle (Gross)Space* of galaxies and stars, as hypothesized by this author, from mysticism point of view, that the process of churning out the galaxies and stars to the *Least Subtle space* happen with the help of the *repulsion force* of the Dark Energy; and Electro-Magnetism is more predominant than Gravity force, in the Dark Energy.

Origin

In 1979, Abdus Salam, Sheldon Glashow and Steven Weinberg were awarded Nobel Prize for contributions, explaining the unification of the Electro-Magnetic Force and the Weak Nuclear force, giving rise to the new terminology of Electro-Weak Force.

The Electro-Weak Energy is important in explaining how all things in the Universe are formed. Steven Weinberg reported that the Higgs-Boson particle is the missing link in the interaction between Electro-Magnetism and the Weak Nuclear Force in which the Weak Nuclear Force causes the Beta radioactive decay of the nucleus, making the neutron split into positively charged proton and the negatively charged electron; and pushing the electron outside the nucleus. With the electrons orbiting the nuclei, the atoms are formed. The atoms combine to form the molecules; and cells and other structures of the Universe are formed from the atoms and the molecules.

Electro-Magnetism and the weak nuclear force are combined to form the 'Electro-Weak Force'. Attempts are now being made to combine the 'Electro-Weak Force' and the 'strong nuclear force' to name them together as 'Electro-Strong Force or Electro-Magnetic Strong Force', as suggested by GUTS- Grand Unifying Theories-, one of the unifying theories of all fundamental energy forces. That will, once again bring the fundamental forces of energy, as suggested by mysticism, described earlier as Gravity force (Ākarshana Sakti) and Magnetic force (Kānta Sakti), to a total of two!

1. Gravity and
2. Electro-Magnetic-Strong Force.

Section Six:

MIND MANAGEMENT

CHAPTER 1

THE BODY-MIND CONNECTION

> I wanted to live deep and suck out all the marrow of life, to live so sturdily and Sparten-like as to put to rout all that was not life, to cut a broad swath and shave close, to drive life into a corner, and reduce it to its lowest terms, and, if it proved to be mean, why then to get the whole and genuine meanness of it, and publish its meanness to the world; or if it were sublime, to know it by experience, and be able to give a true account of it in my next excursion.—*Walden*
>
> — HENRY DAVID THOREAU 1817–1862

THE INFLUENCE OF MIND OVER body has been well documented as *"the mind-body connection"*. On the flip side of the coin, is the influence of the bodily movements and certain poses over the brain and its chemistry; and the resultant changes in the mind. I call this the *Body-Mind Connection*.

It has been discovered recently that neurotransmitters send messages through trillions of synapses across billions of neurons. The neurons are the nerve cells of the body and are responsible for the various functions of the brain and the mind. It is estimated that there are 100 billion neurons in the human body. The influence of the neurotransmitters across the neurons, makes the delineation between body and mind, less and less remarkable. They allow the body and mind to function as a unified system for all practical purposes.

In the 19th century in 1884, William James, an American philosopher published a controversial article that raised the probability that physical actions can elicit corresponding emotion; rather than the emotion causing the physical action, as generally believed as the 'mind-body connection'. For example, when a person runs away from a bear, the immediate reaction of *running away* causes the emotion of *'fear'*, rather than the fear causing the physical reaction of running. It is not *'freight and flight'*, as is commonly believed; it is *flight and freight*.

The emotion of fear sets in after the body reacts to the situation of the bear, based on the knowledge that a bear can attack and harm. The knowledge comes from the memory in the brain. It further proves that mind and body act together as one unit and are not separate.

It raised a controversy that lasted through the next century. It was not until the late 20th century and early 21st century, further research and publications have shown the role of physical poses and bodily movements in changing brain chemistry and emotions. It likely coincided with the introduction and popularity of *yoga practice* in the Western Hemisphere; and studies involving the yogic system are being reported in increasing numbers, proving the changes in the hormones and biochemistry of the brain do occur after *Yoga* practice and sustained bodily movements.

These studies show the benefits of yoga in the neuroendocrine system at the biochemical level and anatomical level.

Recent Studies on the Body-Mind Connection

Using magnetic resonance spectroscopic imaging, Chris Streeter, MD, an assistant professor of psychiatry and neurology at Boston University School of Medicine (BUSM) and Harvard affiliated McLean Hospital, reported that practicing yoga may elevate levels of brain gamma amino butyric acid (GABA), the brain's primary inhibitory neurotransmitter (published in the May 2010 issue of the Journal of Alternative and Complementary Medicine).

This is the first study to show an association between Yoga practice, increased GABA levels and decreased *anxiety*. The findings also suggest that the practice of Yoga can be considered as a possible treatment or as an adjunct to the treatment for depression and anxiety. The World Health Organization reports that mental illness constitutes up to fifteen percent of diseases in the world. Such depression and anxiety are typically associated with low levels of GABA and are often treated with pharmaceutical agents that increase GABA levels.

In the methodology of Streeter's study, two randomized groups of healthy subjects were followed for a twelve-week period. One group practiced Yoga three times a week for one hour per session; the other group walked for the same period of time. At the end of twelve weeks, the researchers compared the GABA levels of both groups, before and after their final 60-minute session. Each subject was also asked to assess his or her psychological state at several points throughout the study.

Those who practiced Yoga reported a more significant decrease in anxiety and greater improvement in mood than those who walked. Over time, improvement in mood and decrease in anxiety, were also associated with 'climbing GABA levels'.

In 2010, Dana Carney and others from Columbia University and Harvard University, published an article in Psychological Science, showing that testosterone and cortisol levels changed, when certain poses were maintained for a total of two minutes.

In this study, 42 subjects—26 females and 16 males— were randomly assigned to the 'high-power' or low-power' poses. The subjects were told that the study was about the science of physiological recordings and was focused on how placement of electrocardiography electrodes above and below the heart could influence data collection, to avoid the influence of subjective feelings on the results, namely the hormone levels.

The subjects' bodies were then posed by a researcher into high power and low power poses. Each participant held two poses for one minute each (a total of two minutes).

The subjects' risk taking was measured with a gambling task; feelings of power were measured with self-reports.

Saliva samples were taken before and approximately seventeen minutes after the power poses manipulation, to measure cortisol and testosterone levels. The poses were harvested from the lexicon of non-verbal communication.

In the 'high power' pose, the subjects were instructed to keep the feet up on the desk with their fingers intertwined behind their heads (an expansive pose, taking more space) for one minute. They were then told to keep their hands on the table with legs spread out (another expansive pose) for one minute.

In the 'low power' pose, the subjects were instructed to sit in a chair, with legs together and hands held together on the thighs (a contractive pose, taking less space) for one minute. They were then told to stand with legs crossed and arms crossed over each other for one minute.

The study was done in the afternoon to control for diurnal rhythms in the hormone levels.

The results showed that *posing in high power* displays caused elevation of testosterone, a dominant hormone; reduction of the stress hormone cortisol; and increases in behaviorally demonstrated risk tolerance and feelings of power. *Posing in low power* displays caused the opposite of the above findings.

These findings suggest, according to the author of this study, 'that the effects of embodiment extend beyond emotion and cognition, to physiology and subsequent behavioral choice'.

Kirk Erickson, a neuroscientist from the University of Pittsburgh and others from the University of Illinois found that people 60 to 79 years old who completed a six-month program of walking *briskly* on a regular basis showed an *increase in the size of the hippocampus and levels of BDNF–brain derived neurotrophic factor*—comparable to those normally found in people almost two years younger.

The hippocampus normally shrinks by 0.5 percent each year, starting at age 40.

There were two control groups in the study– one group did 'toning and stretching' exercises and another did nothing; the two control groups showed no brain changes.

The above three studies provide objective data at the biochemical level and anatomical level, and provide examples of the *Body-Mind Connection*.

CHAPTER 2

THE BRAIN-MIND CONNECTION

*It is a man's own mind, not his enemy
or foe, That lures him to evil ways.*

— GAUTAMA BUDDHA

MIND YOUR MIND

THE MIND IS THE FUNCTIONAL aspect of the brain. If brain is understood as the hardware, the mind can be considered as the software of a computer. However, mind is more than a computer because of the *psycho-neuro-endocrine system with the feedback mechanism* maintaining constancy of *internal environment* and also because of the *limbic system* that is responsible for emotions. Moreover, mind also exists outside the brain

Now understanding the brain and mind is important for the purpose of anger management, anxiety management and general well-being and will be described here in detail in this chapter.

Mind can be a friend or foe. As a foe, it can bring about anger, anxiety, arrogance, jealousy, fear or depression. As a friend, the mind can bring equanimity and balance to life and can be a guiding light. Understanding the mind gives *clarity* and will keep the mind as a friend and guide; and provide health and happiness in life.

Understanding the brain and mind as instruments, similar to other parts of the body will also help to use the mind as an instrument. For example, the hand is the part of the body and it helps us to pick up objects, to eat, to write, and many other functions as an instrument of the body. Similarly, the brain is a part of the body and plays an important role in the functioning of the mind. Brain receives data through the *neurons and neuro-transmitters* from the senses of hearing, touching, seeing, tasting, and smelling by the interaction of the various sense organs with the objects outside, through the medium of mind. The brain sends out impulses for motor functions to various parts of the body. So, mind acts as the facilitator, functioning both inside the brain and outside the brain; and is the functioning aspect of the brain. In fact, the whole body including the brain and mind can be viewed as an *instrument* at our disposal.

The individualized mind, as a component of the nervous system, is like any other system of the body.

For example, heart as a part of the cardiovascular system collects the blood from the body through the veins, sends the blood to the lungs to be purified, for the exchange of CO_2 to O_2, collects the blood and sends it to the rest of the body. Similarly, the individualized mind facilitates the brain to be a relay station; and receives the data through the neurotransmitters and sends it to different parts of the brain and initiates the appropriate action, needed.

So, a brief explanation is given below about how the brain and mind work, as applicable to our present discussion of how the mind works and to the mind management.

THE HUMAN BRAIN

The human brain and nervous system contain over one hundred billion nerve cells, known as *neurons* that communicate with connections, known as *synapses* which number in trillions. Even though the human brain has the same structures as that of other mammals, it is relatively larger in size. It is also, much more specialized and unique in the cerebral cortex with higher reasoning capacity and is able to learn and speak languages. The brain consists of

1. the cerebrum
2. the cerebellum
3. the brain stem

The Cerebral Cortex

The cerebrum has the cerebral cortex as the outermost layer of the cerebrum and consists of four lobes on each side. The cerebral cortex has multiple deep folds, known as gyrus (gyri) and sulcus (sulci), thus increasing the surface area many-fold (no pun intended). The lobes of the cerebral cortex are

1. The frontal lobe
2. The parietal lobe
3. The temporal lobe
4. The occipital lobe

The functions of the right side of the body are controlled by the left side of the brain and the left side of the body are controlled by the right side of the brain.

The frontal lobe is situated in the front of the brain, as the name implies; and is responsible for the functions of reasoning, language, higher level of cognition and motor skills. The motor cortex is located at the back of the frontal lobe and provides voluntary movements as needed.

The parietal lobe is situated behind the frontal lobe and lodges the somatosensory-cortex which is responsible for the processing of the senses, such as touch, pain and pressure.

The temporal lobe is in the lower section of the brain and contains the *auditory cortex* that is needed for *hearing*. The hippocampus, a part of the limbic system is situated in the temporal lobe and is responsible for the function of *memories*.

The occipital lobe is in the back portion of the brain and contains the *visual cortex* that is essential for *seeing* and receives information from the retinas that form the back of the eyes. The right sides of both retinae are controlled by the left visual cortex; and the left sides of both retinae are controlled by the right visual cortex.

THE CEREBELLUM

The cerebellum is situated behind the brain stem and below the cerebrum. Even though it is relatively smaller in size, it contains more than 50 percent of the total neurons of the brain, i.e., nearly fifty billion neurons. It helps to control posture, balance and the coordination of voluntary movements.

THE BRAIN STEM

The brain stem consists of

1. hindbrain and
2. midbrain.

1. The hindbrain consists of medulla, reticular formation and pons.

The *medulla* or medulla oblongata connects the brain to the spinal cord and controls the autonomic functions, heart rate, blood pressure and breathing.

The *reticular formation* is a network of nerves in the medulla and helps to control functions of sleep and attention.

The *pons* is located between the medulla and cerebrum and helps to coordinate movement on each side of the body.

2. The Midbrain is the smallest structure of the brain. The midbrain contains a cluster of structures called the *basal ganglia*. It coordinates messages between multiple areas of the brain. It controls visual and auditory systems as well as eye movement.

The part of the midbrain that contains *substantia nigra* and *red nucleus* controls the body movement. The substantia nigra contains a large number of dopamine producing neurons; the degeneration of these neurons causes Parkinson's disease.

THE MIND

The mind is the functional part of the brain and collects data from its surroundings, by way of the five senses and the corresponding sense organs of ears, skin, eyes, tongue and nose.

The *Thalamus* acts as a relay station, receiving messages from all the sensory receptors of the body and sending the information to the different regions of the brain, as described earlier. These data are either used in the present moment or stored in the memory as past events. The mind also is able to analyze the data and come to conclusions.

THE LIMBIC SYSTEM AND EMOTIONS

The emotional aspect of the mind functions through the limbic system of the brain. The limbic system includes various structures, such as *hippocampus, amygdala, limbic cortex and the septal area* which form connections to the hypothalamus, thalamus and cerebral cortex.

These structures are involved in emotions, learning and memory. The Limbic system operates by influencing

1. The endocrine system and
2. The autonomic nervous system.

It acts on the endocrine system through the *hypothalamus*, which regulates all the endocrine glands by way of various releasing factors, which regulate the various *trophic hormones from the pituitary gland*; e.g., thyrotrophin-releasing-factor from the hypothalamus regulates the thyrotrophin secretion from the pituitary gland; which in turn regulates the thyroid hormones from the thyroid gland.

Similarly, various trophic (stimulating) hormones are secreted by the pituitary gland, such as *adreno-corticotrophic hormone* regulating *cortisol* from the adrenal cortex, and *gonadotrophin hormone* that acts on ovaries and testes, etc. The pituitary gland is therefore aptly called the *master of the orchestra*.

Oxytocin, another hormone produced in the hypothalamus and stored in the posterior pituitary gland is given a nick-name of 'love-hormone' or bonding hormone. Recent studies associate this hormone in various bonding relationships in both men and women, such as orgasm; and especially in women both during and after childbirth. Supposedly, it is released in large amounts in child birth during the dilation of the uterine cervix; and helping the labor and promoting maternal bonding and lactation after child birth. This hormone is also used during childbirth as a 'Pitocin-drip', for uterine contraction, if needed.

The other hormone from the posterior pituitary is called *Vasopressin or Anti-diuretic hormone* (ADH). As the name implies, it causes an increase in blood pressure, by water retention and contraction of arterioles. Reduced secretion of ADH causes Diabetes Insipidus, by acting on the tubules of the kidneys to produce excessive dilute urine, which results in dehydration.

Male aggression is reported to be associated with this hormone.

The limbic system also influences the *autonomic nervous system* which consists of the *sympathetic system* and the *para-sympathetic system*.

Epinephrine and nor-epinephrine are the hormones of the sympathetic system, secreted by the adrenal medulla. *Acetyl-choline* is the hormone of the para-sympathetic system.

Anger is a reaction to a situation either self-created or created by others. Anger expresses itself through the sympathetic system by secreting more *adrenaline* from adrenal medulla and sympathetic nerve endings; and by secreting more *cortisol* from adrenal cortex.

Anger and its association with the adrenaline response, leading to increased heart rate and stress response with increased cortisol are well documented.

Whatever situation or person is the source of the anger, the person *reacting* with anger is the only one who can manage the *reaction*, i.e., by *acting* instead of reacting to the situation. The method of management is further explained under 'anger management'.

Parasympathetic stimulation, in general, causes relaxation and lessening of anxiety. This is made use of, in many of the yoga poses and pranayama breathing that stimulate the *parasympathetic nerve*, *Vagus* nerve. The Vagus is one of the *cranial nerves* (tenth cranial nerve). It is named as the vagus nerve, because it takes a meandering and long course all over, like a vagabond, from the brain all the way down to the abdomen, giving out multiple branches along the way in the head, neck, thorax and abdomen. Besides its motor and sensory functions, it is a parasympathetic nerve and thus helps in lowering anxiety and improving relaxation of the mind.

Neurotransmitters

Neurotransmitters are chemicals that transmit messages across the synapses of neurons (the nerve cells). The human body consists of about 100 billion neurons. Some of the important neurotransmitters are Epinephrine, Acetylcholine, Cortisol, Dopamine, Melatonin, Serotonin, GABA–gamma amino butyric acid–and Endorphin.

Thus, the various structures of the Limbic system, with varying degrees of emotional attachment and memory of past events, form a part of the network of the psycho-neuro-endocrine system.

Memory and Anger

The limbic system, briefly explained above, is involved in Emotions, such as anger, and also in memory. The Cerebral Cortex also plays a part in memory. The past events are stored in the brain as memory. Very often, these stored *past* events from memory are the trigger for anger in a person.

Future planning of a successful outcome or ambition along with *emotional attachment* also can be a trigger for the onset of anger. For example, in the game of tennis, with a game score of 6-6 and a 'tie-breaker' score of 6-5, the serving player is unable to hold the point to win the set-point for a successful outcome and eventually loses the game and/or the match; and this failure of the planned successful outcome can be a trigger, depending on the degree of *emotional attachment* in that situation.

The mind and body are not independent of each other and can be considered as one entity. For example, the brain is an anatomical part of the body and the mind is one of the physiological functions of the brain.

The brain and nervous system, as alluded to earlier, contain over one hundred billion neurons which are all connected with trillions of synapses which send messages.

Recent studies also reveal that the neurons produce many hormone-like neurotransmitters, such as dopamine, serotonin, endorphin and gamma amino butyric acid (GABA) that influence the mind. It makes us wonder if the brain can be a very large and complex endocrine gland; on the other hand, if all the endocrine glands with their hormones, can be considered part of the brain and the nervous system.

When the mind goes into anger mode, the body reacts with increased secretion of adrenaline hormone (aka epinephrine) from the medulla part of the adrenal glands, similar to *fight or flight* response or survival response.

This increases the pulse rate and blood pressure; the eyes become red.

Similarly, the mind reacts to the occurrences of the body. For example, if the finger goes near fire accidentally, there is pain—the sensory nerves take the message from the affected finger to the brain, to the area of touch, pain and temperature; and the brain reacts instantly and sends the message through the motor nerves to the corresponding arm muscles to pull back the hand quickly. All of these happen in a split second. In fact, the pain is a warning sign, given by the anatomical brain to the finger, through the physiological pain sensation. The nerve to the brain is a sensory nerve and the nerve from the brain to the arm muscles is considered as a motor nerve. If the fire happens to be severe, resulting in burns to the area, the emotional part of the mind also comes into play and shows further disappointment and anger. If the fire was caused by another person, then the mind may react with anger towards that person.

Similarly, when there is *'itis'* or inflammation of any part of the body, e.g., appendicitis, there is pain which is a warning sign. But if it is not treated properly, the mind may show anger and disappointment.

Depression After Heart Attack

Another example where the mind reacts to the ailment of the body, is heart attack.

Even after the proper treatment of the cardiac episode, presenting with severe pain, it is not uncommon for the patient to go into *depression* with feelings of self-pity and the mind questioning why *it* had to happen to him. Besides depression, *insomnia* also is a common occurrence, after heart attack, needing sleeping pill for adequate rest.

Ego

Past events along with future plans and ambitions give the mind one's personality or ego. Thus, the personality is considered a combination of the past and future; and not the present moment of awareness. This is how the anger episodes are the result of, and a part of one's personality; and not a reflection of the *Awareness of the present moment*.

Here again, in the practice of *'Awareness of the present moment'*, the emphasis should be on *Awareness* and not in the *present moment*. Many of the teachings that stress upon the *present moment* have failed because of the misguided emphasis on the *present moment*, instead of on *Awareness*. The reason is 'the present moment' is still a part of *time*, as opposed to *past and future*. *Whereas, Awareness is beyond time.*

This Awareness is more evident in the higher intelligence of human beings. Thus, human beings do not have to react on instinct, with anger, for example, like animals. The *Awareness* will be further explained in detail below.

AWARENESS
Human beings are intelligent enough to be aware of one's own mental process and thoughts. The word, "thought" itself is the past tense of the word, "think". So, many thoughts are of past events or past thinking. Awareness has the capacity to *be aware* of all things that are based in time and space, especially in the present moment. Thoughts are time based and mostly, pertain to the past. Thus, Awareness provides the capacity to be aware of *one's own thoughts*, as and when they occur; in other words, to be aware of the mind itself.

Awareness is also called consciousness sometimes, creating confusion because consciousness generally pertains to consciousness of the mind as opposed to unconsciousness; whereas Awareness, in this context, pertains to *being aware of the mind* itself and is *beyond mind*. That is the reason the word, *Awareness* is being used here, instead of the word, consciousness.

CHAPTER 3

MIND MANAGEMENT

> But if life itself is good and pleasant (...) and if one who sees is conscious that he sees, one who hears that he hears, one who walks that he walks and similarly for all the other human activities there is a faculty that is conscious of their exercise, so that whenever we perceive, we are conscious that we perceive, and whenever we think, we are conscious that we think, and to be conscious that we are perceiving or thinking is to be conscious that we exist... (*Nicomachean Ethics*, 1170a25 ff.)
>
> — ARISTOTLE

EVEN THOUGH THE MIND IS used as an instrument in the understanding of Awareness, the *Limitless*

Awareness cannot be contained in the limited mind. The mind is similar to a stick that is used to kindle a *camp-fire* which consumes the stick; and the camp-fire remains. Sri Ramakrishna, a *Mystic* who lived in the 19th century, compared the mind to the stick that is used to kindle the *funeral* pyre; and the stick i.e., the mind is consumed in the fire and the fire remains, representing the realization of *Self*, the ultimate goal of Vedanta philosophy which Sri Ramakrishna was teaching.

The mind can be contained in the Awareness; i.e., the Awareness can be aware of the mind, as a flow of thoughts. But Awareness exists even when the mind is turned off, as in deep sleep.

If we close the eyes and pick up a cup with the hand, we realize that act; i.e., we are aware of the function of the hand, through the senses. By closing the eyes, the seeing-sense is removed; but the tactile sense of the finger, touching the cup is present. However, in a person with an artificial arm, for example, the tactile sense and the motor function are removed. So, with the eyes closed, in this person, the sense of seeing and touching, and the part of the brain that collects the sensory data, are not involved in the act of picking up the cup.

Perhaps here, the higher intelligence that can be *aware* of the functioning of the mind and a part of the function of the mind to activate the artificial limb are what are behind the awareness of the act of picking up the cup.

Let us take another example. When we are asleep, the mind is turned off, not unlike putting the computer to sleep-mode. *The Life-Energy along with the Inherent Awareness* keeps the brain and the mind alive, along with memory and along with the data of who we are, even though the mind is completely turned off, in deep sleep. When we wake up, the mind becomes alive and awake; and has no question about our identity and proceeds with regular functioning of the body and mind.

The Mind itself is not unlike a super-computer. A computer has hardware and software and the data are stored and maintained by the computer's energy. If we consider the brain as the hardware, then the mind is the software. However, the addition of *limbic system* and the associated *emotions* and the *psycho-neuro-endocrine system* with *feedback mechanism* regulating the *internal environment* distinguish the mind and brain from a computer.

Brain, the hardware, as described earlier, has the cerebral cortex, cerebellum, brain stem with the midbrain along with the multiple structures of hippocampus, amygdala, thalamus and the hypothalamus together called the limbic system, connecting to an elaborate network of 100 billion neurons, through neurotransmitters and hormones.

The mind, the software, functions with analytical, logical and creative capabilities with sensory receptors and motor functions; and through the limbic system, is the seat of emotions, learning and memory.

MILELA THEORY AND MIND

Milela-Theory (Mutually Inherent Life Energy and Limitless Awareness) has been dealt with in detail in the beginning chapters of this book. The connection between the all-pervasive Awareness-Intelligence and mind is further explained here.

As explained earlier, in the chapter on *Milela*, we see babies born every day and people passing away all the time. However, we take the Life-Energy that comes and goes with birth and death for granted. Whereas, Life-Energy is the very basis of our existence.

Quantum mechanics also has shown the atoms and the subatomic particles are the basic unit of life; The Special Theory of relativity of Einstein has given the equation $E=Mc^2$ which essentially says energy(E) and mass(M) are different forms of the same thing; c is the constant of the speed of light (186,000 miles per second). This energy is the Life-Energy of everything that is all-pervasive including our body and mind.

Milela-theory says Limitless Awareness is inherent in the Life-Energy.

This Inherent-Awareness-Life Energy is present as the *substratum* of the human mind as well as the world, we live in, and in the cosmos, with the precise functioning of planets and stars and billions of galaxies and multiple universes, in spite of any *chaos in the cosmos as well as in the human mind.*

What is unique for human beings is that the higher Awareness-Intelligence of the humans is also capable of being *aware of the mind* itself, as a flow of thoughts, with one thought connecting to the next, like a chain.

THE TECHNIQUE OF MIND MANAGEMENT— *ANGER MANAGEMENT–CHETANA YOGA OR CHAITANYA YOGA*

Chetana or Chaitanya means Awareness. In the technique of anger management, the above described Awareness can be effectively used as follows:

This awareness capability of the human intelligence can also see one's own mind as thoughts. Just like we can see the hand with our eyes, we can learn to see our own thoughts; i.e., to be aware of the thought. So, the technique is to see the thoughts— in other words, to be aware of the thoughts— as and when the thoughts occur during the day.

This technique has also been explained as *Chetana Yoga* or *Chaitanya Yoga* in the beginning of this book. Seeing the thoughts has to be developed as a *habit*, during the *normal activities of the day*, when there is no anger.

To further clarify, human beings have the capacity to realize this Awareness; i.e., to be aware of the Awareness. Developmentally, human beings are at a more mature level than animals, to *be aware* of this Awareness and the mind, as a flow of thoughts, whereas animals have the capacity to think and act on instinct only. That is the difference between human beings and animals.

So, when men or women act on instinct only, the animal nature of mind predominates. On the other hand, when men or women act with *kindness and good intentions* in thinking followed by the same attitude in all communications, including speech and modern social media, leading to actions with good intentions, the divine nature of the mind predominates. Such actions will not produce *fear* in the mind. Fear is the cause of *anxiety*.

Such actions will also prevent situations of anger. If the situations of anger arise, the *Chetana Yoga habit* will help in evaluating the thoughts and acting appropriately; and will prevent the *actions of anger-reaction*. In other words, *being aware habit* has created a permanent *layer* between the *reactionary mind and actions of anger-reaction*. Instead, the person is able to do the appropriate actions needed, without reacting with anger outbursts.

This attitude of *Kindness and Good intentions* will also help in all relationships, such as husband and wife, parents and children, brothers and sisters, friends and contacts, employer and employee etc.

Animals and birds also have the Divine nature of Kindness and Good intentions, in varying degrees. For example, even wild animals take care of their offspring providing food and with caring affection. Birds bring food in their mouth and feed their babies. When the babies are old enough to fly, the mother bird nudges them out. We have all seen pet dogs that are tolerant and kind to our children, even when they sit on the back of the dog.

Humans, however, have a higher intelligence and the ability to be aware of the flow of thoughts and the ability to choose. The well-known psychologist, Carl Jung (1875-1961CE) developed his system of psycho-analysis and psycho-therapy based on this awareness-intelligence, also known as *consciousness*

Human beings are a complex combination of *animal nature and divine nature*. In our daily life, we all have a predominance of either divine nature or animal nature. When the animal nature takes over, generally speaking, we react to circumstances, instinctively.

This results in outbursts of anger or feelings of jealousy or attitudes of arrogance or vanity or greed. All of these have a common thread of *reacting instinctively* to the circumstances; and anger tops them all because the outbursts of anger are uncontrollable for that person, at that moment.

That is the reason this technique of seeing thoughts is recommended and then the person can *be aware* of the thoughts as they occur and then it becomes a habit in due course. However, trying to be aware of the reaction of anger, *only* at the time of anger is not possible, unless one cultivates a habit of seeing the reactions of one's own mind, as they occur, in *less severe circumstances*. Cultivation of this habit is the key to anger management.

LONELINESS AND DEPRESSION MANAGEMENT
CHETANA YOGA—A REMEDY FOR ISOLATION

Some seek solitude and choose to be alone; and that is not loneliness. So, it is not the physical solitude; but it is the state of the mind. Loneliness is *the subjective feeling of isolation from friends and family*.

AMILEA is introduced here, at this point, as an acronym for *All-pervasive Mutually Inherent Life Energy and Awareness*, to emphasize the *all-pervasiveness of Awareness inherent in the Life-Energy*, including our Life-force, our true nature. It is the same as *MILELA* that has been discussed in detail, in the chapter under the same name.

The practice of Chetana Yoga, based on the theory of AMILEA or MILELA—*emphasizing the all-pervasive Awareness as our true nature or inner self*– leads to a sense of belonging. Moreover, ignorance of this theory of all-pervasiveness of our inner-self leads one to a lack of equanimity.

Along with the understanding of the all-pervasiveness of our inner self, the *Chetana Yoga* of being aware and seeing our own thoughts, as they occur, as a *habit*, creates a sense of equanimity and a sense of belonging, due to the all-pervasiveness of *being aware*. The lack of equanimity and a lack of sense of belonging leads one to *isolation, with a subjective feeling of isolation from others*, and *loneliness* and eventually to *depression*. So, Chetana Yoga, based on the understanding of the all-pervasiveness of our inner self, is recommended to develop the habit of seeing your own thoughts, to prevent as well as to wean the mind away from loneliness

Frustrations of the mind due to various reasons, often leads to the development of *various types of prejudice*; and prejudice and the feeling of isolation lead to loneliness.

An understanding that our inner nature or true self is the All-pervasive Awareness leads one away from the feeling of loneliness. So, the practice of *Chetana Yoga based on the All-pervasiveness of our inner self* is again emphasized as a necessity with the understanding that every one of us is a part of the whole, the *all-pervasive Awareness space*; and there is no justification in prejudice of various kinds, isolation and loneliness.

The practice of Chetana Yoga, described earlier, in detail, in the chapter under the same name, creates also a permanent layer between the thoughts and action, before the action is committed, enabling *equanimity* and balance to the mind. Equanimity or *Samathvam* is Yoga, so says Bhagavad Gita in Sanskrit,

"*Samathvam yogam uchyate.*"

STRESS MANAGEMENT

Hans Selye introduced the term *Stress* in the twentieth century and proved that the hormone cortisol, from the adrenal cortex, is increased at times of stress.

Stress is a part of life and cannot be totally avoided. After one accepts the inevitability of stressful situations coming and going, one has to learn how to cope with stress; therein lies the prevention of damage to our mind and life.

People often say that a person externalizes his or her reaction to stress by acting out one's emotions of frustration and/or anger. In the same way, we say that a person internalizes his or her reactions to various situations, if one does not show the emotions outside.

Most people *wrongly* assume that externalizing a stress reaction is better, because the common concept is that internalization leads to suppression of emotions in the subconscious mind, causing inevitable damage to the psyche, resulting in damage to the body as well; and at some future time, the welled-up emotions might burst out.

The truth is, the above described internalization and the externalization concepts pertain to the emotional and mental make-up of *reactions of the mind* in general. So, both of them cause equal amount of damage to the mind and body of that person, not to mention the ramifications involved in various relationships in different situations of stress.

So, the internalization and externalization of reactions to stress do not appear to be the proper approach to managing stress.

Then what is the proper approach to stress management? First of all, one needs to comprehend and view internalization and externalization as products of the mind and its reactions; they both need to be viewed as external; and together they are to be viewed as a *mental response*.

In order to successfully manage stress, one has to take on a *Supra-mental approach*. In the supra-mental approach, one views the situation, as though from a gallery above, that is playing out the different roles of a drama; this way one can assess the situation quickly without reacting. Hence one will be able to come up with a solution for the immediate moment as well as for the long-term.

Here again the body and mind are together viewed as an instrument- the body being the *outer instrument* and the mind being the *inner instrument*.

Practice of *Chetana Yoga* or *Chaitanya Yoga* helps one to have a *supra-mental* approach for managing stress. As described earlier, *Chaitanya Yoga* involves *being aware* of the thoughts and mental responses as and when they occur under normal circumstances during the day when there are no moments of *stress*.

ANXIETY MANAGEMENT

When men or women act on instinct only, the animal nature predominates. On the other hand, when men or women act with *kindness and good intentions* in thinking followed by the same attitude in all communications, including speech and modern social media, leading to actions with good intentions, the divine nature predominates. Such actions will not produce *fear* in the mind. *Fear is the cause of anxiety.*

Anxiety, on the tennis court for example, expresses itself as choking; it often comes up when the player is winning, for example, with a score of 5 games to 2 and the player has to get just one more game to win the match. When anxiety takes over, attention to playing the point is diverted to thoughts of winning the sixth game and finishing the set; this thought process keeps the anxiety level up and takes away attention from the ball and the movement of the body that are needed to execute the shots.

So, the antidote is to *pay attention* again to each point, focusing on the ball and executing the necessary movement of the body and the racquet, which is like an extension of the body at that moment.

This is true in life also:

Let us examine the nature of anxiety in a little more depth; anxiety is usually about the *fear of the result of the action* that is taking place, the effect that is going to happen as a *result of the action or fear of a past event.* On close scrutiny, we find that we only have *control* over what we are doing; not over the result. This has been analyzed and explained in Bhagavad Gita, an ancient text that explains how to perform our duty or any action in our daily life, without anxiety— as an advice given by Sri Krishna to Arjuna in the battlefield— metaphorically in the *battlefield of our mind.*

So, the technique to avoid anxiety is *to focus on the action itself;* and not on what is going to happen or what might happen; in other words, not on the results of the action. It does not mean, we should not consider the results of any action, before executing the action. It only means, we should *focus* on the action only, at the time of action. Because, focusing on the results of the action will, invariably, take the attention away from the action.

From the above discussion, we see that *fear* causes anxiety. So, practice of *kindness and good intentions in thinking* also, followed by the same attitude in all *communications, including speech and modern social media*, leading to *actions with good intentions* will not produce *fear* in the mind, thus preventing *anxiety*.

Anxiety can also be triggered by recollection of past events in one's life. Even though it is very obvious that past is dead, a person holds on to the past in his memory bank and brings it up to establish itself. Again, we find that the personality exists as the combination of past and future, namely past events stored in the memory and future ambitions and wishes. In order to avoid letting past events and memories cloud the present, the technique, again, is to focus on the action itself, with full attention or awareness in the present moment, avoiding onset of fear.

> "Dead yesterday, Unborn tomorrow
> Why fret about them,
> If today be sweet?"
>
> –Omar Khayyam, Sufi poet

CRISIS MANAGEMENT

The inspiration for the following poem on 'Crisis Management' was based on a phone call from one who did not believe in the power of Chetana yoga and being aware of the mind. Nevertheless he was desperate due to the thoughts of suicide and homicide; and was seeking immediate help and reluctantly tried my advice, narrated in this poem. He was able to come out of the 'crisis.'

CRISIS MANAGEMENT

When waves of crises bombard the mind,
Even thoughts of suicide and homicide,
Pay attention to the chain of thoughts;
Observe continuously again and again.
See the thoughts of the crises
Lose their power and turn powerless
Slowly but surely with Awareness,
Mind then becomes calm and rational,
Find a solution to the root-cause of the crisis.
If it is 'attachment' to a loved one,
'Let Go' the attachment,
For love is not attachment,
Attachment is possessiveness, whereas,
Children and other loved-ones
Are not our possessions.
For attachment leads to 'control and violence'
Love is affection and acceptance,
Mutual respect and caring.
If it is events of the past, 'accept',
Accept, whatever it amay be,
For you cannot change the past.
Stop building expectations,
For future is really not in our control.

Section Seven:

MEDITATION

CHAPTER 1

PRANAYAMA

❖ ❖ ❖

The mind is the king of the senses
and *Prana* is the king of the mind.

— HATHA YOGA PRADHIBIKA

PRANAYAMA, A SANSKRIT WORD, HAS different connotations. Prana literally means the life-force or life-principle. It also implies the breathing itself. Yama means control. So, through the control of the breathing process, *pranayama controls the mind*.

So, before going into the details of pranayama, it is important to understand the basic nature of mind. A brief description of the mind is given here.

The function of the mind is to gather information through the *senses* of hearing through the ears; of touch, pain and temperature through the skin; of seeing through the eyes; of taste though the tongue; and smell though the nose.

The mind also mediates speech, the motor functions of the upper and lower extremities, and the actions of excretion and procreation. Animals also use the mind to gather the above five senses and to perform the above five motor functions.

However, the intelligence to analyze, discuss and deliberate is of much higher caliber in human beings. Human beings also have *free-will*. Perhaps, for the same reason, humans have stronger egos.

Emotions also play a role in the functioning of the mind. The right side of the body is controlled by the left side of the brain and is called the dominant side— for a right-handed person— and vice versa.

Generally, the left side of the brain is responsible for analytical and logical thinking and the right brain is considered to be creative and intuitive. However, the right and left sides cross over in the middle of the brain.

As we discussed before, the mind, namely the flow of thoughts, can only be regulated by either approaching it from *below the level of the mind* through the different breathing techniques of Pranayama; or from *above the mind*, by paying attention to the mind and thus being aware of thoughts as and when they occur, a habit described as *Chetana Yoga* or *Chaitanya Yoga* in the beginning of this book. This habit regulates the flow of thoughts and eventually goes beyond the mind, described earlier as the *supra-mental approach*. Meditation also helps to reinforce *Chetana Yoga* to some extent, by emptying out the *cobwebs of thoughts*, the mind had woven in the past.

PRANAYAMA:

Pranayama is helpful because the breathing is deliberate, and also helps in the control of mind. The different techniques of Pranayama are described below.

Before doing pranayama, traditionally, one should drink three sips of water.

There is an important reason behind this ritual, namely to prevent aspiration of food-particles that may be lodged in the throat.

Yogic Breathing–Deep Breathing

Yogic Breathing is a type of Pranayama and can be practiced at any time or place, and without much training or guidance.

The technique is as follows:
Inhalation:

There are three phases in the inhalation.

1. In the first phase of breathing in, the *abdomen slightly pushes out*, filling the lower, wider parts of the lungs.

2. In the second phase of inhalation, when the middle parts of the lungs are filled with air, the chest pushes out. When the chest expands, the *abdomen tends to pull in*; it is perfectly normal and it is not advisable to keep the abdomen pushed out.
3. In the third phase, the collar bones go up, when the *shoulders are pulled back and up*, expanding the chest further, filling the apical or upper parts of the lungs.

Exhalation:
For breathing out, the steps are reversed; namely the shoulders drop down; the chest and abdomen relax.

This type of deep breathing stimulates the *vagus nerve* which is a parasympathetic nerve; thus, this technique brings about relaxation and reduces anxiety and lowers the heart rate.

Pranayama Methods
In the Pranayama method, there are three phases:

1. Puraka
2. Rechaka
3. Kumbhaka

Puraka means breathing in and Rechaka means breathing out and Kumbhaka means retention.

This is generally done in the proportion of 1: 4: 2 for inhalation, retention, and exhalation, respectively. However, whenever Kumbakha or retention of breath is included in the practice, this is considered an *advanced Pranayama* and it is advisable to seek guidance from a teacher who is familiar with this method.

In the beginning, the retention is to be started at 1: 1: 1 for inhalation, retention and exhalation; and one may slowly build the retention up to 4. So, the 1:4:2 ratio should not be practiced in the beginning; it should be 1:1:1, and gradually increased. This ratio of 1:4:2 is only mentioned here as the traditional system of practice.

Once you master the 1:1:1 ratio, you may start the 1:1:2 ratio using the proportion of 1:1:2 for inhalation, retention and exhalation. This will help to increase the duration and intensity of the next 'breathing-in'. To reiterate this point, the exhalation should be done 'slowly' and be longer in duration.

The practice of Pranayama, if one wants to pursue it, should be done carefully and slowly and gradually increased as follows:

Sit comfortably in ArdhaPadmasana (half-lotus pose) or Vajrasana (diamond pose–legs folded back and sitting on the heels of the feet) or Sukhasana (comfortable sitting position) or in a chair, if it is preferred. Use the right hand for closing and opening the nostrils; the thumb for the right nostril and the ring and little fingers for the left nostril.

1. IPSI-LATERAL OR SAME-SIDE PRANAYAMA
Inhalation is through the left nostril and exhalation is through the same nostril

Repeat 6 times; start with 3 times and gradually increase to 6 times.

There is no retention in this Pranayama.

Repeat the above steps through the right nostril.

2. BILATERAL OR BOTH SIDES PRANAYAMA
Inhalation is through both nostrils and exhalation is through both nostrils, slowly.

Repeat 6 times; start with 3 times and gradually increase to 6 times.

There is no retention in this Pranayama This is similar to 'Yogic Breathing'.

3. CONTRALATERAL OR ALTERNATE SIDES PRANAYAMA
This is also known as 'Nadi Sodhana'.

Inhalation is through the left nostril. Exhalation is through the right nostril.

Inhalation is through the same right nostril
Exhalation is through the left nostril
This is one cycle.
Repeat 3 times and gradually increase to 6 times

4. ALTERNATE SIDES—NADI SODHANA PRANAYAMA WITH KUMBHAKA

As mentioned above, this advanced system should be done with guidance from a Yoga teacher.

In the Nadi-Sodhana type of Pranayama, one inhales for one count through the left nostril and holds for the one count and then exhales through the right nostril for twice as long–so the proportion between Puraka and Kumbhaka and Rechaka— inhalation and retention and exhalation— is 1:1:2. One can use either the number or the word "OM." Then the duration is increased gradually, taking care to use the same proportion;

1:1:2
2:2:4
3:3:6
4:4:8 and so on
The method is as follows:
Inhalation is through the left nostril.
Retention.
Exhalation is through the right nostril.
Inhalation is through the right nostril.
Retention.
Exhalation is through the left nostril.

This completes one cycle. This is the Puraka-Kumbhaka type of Nadi Sodhana.

Kumbhaka

In the Kumbhaka or retention phase, the chin-lock or *Jalandhara Bandha* and rectal contraction or *Moola-Bandha* are recommended to be done in that order; and at the end of retention, rectal contraction is relaxed first and then the chin-lock is released, before the exhalation.

The chin-lock is done by bringing the chin to the top of the sternum (the breast bone), with slight flexion of the neck.

After a month of this practice, one may do Rechaka-Kumbhaka as follows:

Inhalation is through the left nostril.
Exhalation is through the right nostril.
Retention.
Inhalation is through the right nostril. Exhalation is through the left nostril.
Retention.

The proportion for the above is 1:1:1 in the beginning. After a month of this practice, the proportion can be increased gradually to 1:2:4 for inhalation, exhalation, and retention respectively, as follows:

1:1:1
1:2:2
1:2:3
1:2:4

It is very important not to increase the proportion prematurely; in fact, it is not important to do any of these methods vigorously and it is more important to do them consistently every day and preferably at the same time of the day.

Next, the above two, namely Purakha-Kumbakha and Rechaka-Kumbakha, i.e., inhalation-retention and exhalation-retention can be combined as follows:

Inhalation
Retention
Exhalation
Retention
Inhalation
Retention
Exhalation,

using the gradual progression ratio, as mentioned above.

So, the emphasis is to progress gradually in Pranayama.

A Word of Caution.

A word of caution is warranted here regarding the retention phase of pranayama, Retention is not advised for anyone with health problems, such as high blood-pressure, cardiac or pulmonary diseases.

If you are starting the practice in the later years of life, retention is not advised. In fact, for any age, a moment of retention both after inhalation and after exhalation is all that is needed; and it is strongly recommended that any pranayama practice that includes retention or Kumbhaka should be done under the guidance of a yoga teacher who is familiar with pranayama.

Specialized Pranayamas

Besides the regular Pranayama described above, there are some specialized Pranayama practices that are recommended with some specific benefits. For example, *Bastrika or bellow-breathing* increases the heat in the body and so it is advisable to practice Bastrika more in the winter season and less in the summer.

Seethali tends to cool the body and so it is beneficial generally in the summer season. However, it also helps to lower blood pressure in patients with high blood pressure; hence, it can be practiced regularly in such circumstances.

It is advisable to seek guidance in learning these techniques.

Bastrika or Bellow Breathing

As mentioned above, Bastrika increases the heat in the body; it also increases the energy in the body and so morning time is best suited for practice of this breathing. It is best not to do Bastrika before bed-time, to avoid sleep disturbances. It is also preferable to have empty stomach or three hours after a meal or at least one hour after drinking any liquid.

One sits comfortably on the floor or in a chair. Place the hands firmly on the knees and take a deep breath in and out; and start the bellow like movement of the abdominal muscles, breathing out when the abdomen is pulled in, and breathing in when the abdomen is pushed out.

The emphasis is on breathing out and pulling the abdomen in. The breathing in and out are of equal proportion and are done with both nostrils.

Start at a comfortable pace and increase the pace faster to about 20 times; then take a deep breath in and slowly breath out, pulling the abdomen in, to complete one cycle. Repeat 3 to 6 times as needed.

Besides increasing body heat as mentioned above, it also helps to prevent allergy symptoms, perhaps by regulating the adrenal hormones. The adrenal glands are situated in the back of the abdominal cavity, above the kidneys. So, the Bhastrika is also advised in the spring season or allergy season, when the pollen count is high, to prevent nasal allergy symptoms.

Kapāla Bhati

Kapala means the head or skull in Sanskrit; bhati means 'shines'. Kapāla-bhati helps one to have a shining face.

It also helps one to prevent nasal allergy symptoms. The method is similar to Bastrika; the main difference is that in Kapāla Bhati, the abdomen is used minimally, and the forceful expirations happen at the nose level in Kapāla Bhati.

Sit comfortably on the floor or in a chair. Similar to Bastrika, the method is to do forceful exhalations about 20 times rapidly; and then take a deep breath in and slowly breathe out. The rapid exhalations are similar to blowing the nose to clear the nasal passages; the exhalations are done with both nostrils. There is only a slight movement of the abdomen with each rapid exhalation.

Vipareeta Svasa.

The method is similar to the above Kapāla-Bhati, except the rapid exhalation-movements are done at the tip of the nose.

Pulling the tongue up against the palate and pulling the chin and abdomen inwards help to close the air-passage and facilitate performance of the Vipareeta svasa. The benefits are similar.

The above three are described together here, to show the similarity and the differences.

Seethali Pranayama

As mentioned earlier, Seethali pranayama tends to cool the body and also to lower the blood pressure.

Sit comfortably on the floor or in a chair. Breathing in is done through the mouth and breathing out is done through the nose, as follows:

Make a tube of the tongue, by folding the tongue, with the tip of the tongue at the lips; and protrude the tube out. Breathe in through the tube of tongue slowly, filling the chest and expanding the chest, filling up the lower lobes of the lungs with air; then pulling the shoulders back, thus filling up the upper lobes of the lungs. Pull the tongue in and close the lips together.

Hold the breath briefly, raising the tongue to the front half of the palate and pulling the abdomen in. Lock the chin to the upper chest (also known as *Jalandhara Bandha or Chin-lock*).

Breathe out slowly, relaxing the chest and abdomen. Repeat three to six times.

UJJAYI PRANAYAMA:

Ujjayi breathing is an important pranayama to learn, as it can often be an integral part of other Yoga practices, such as in preparation for meditation, mantra yoga, kriya yoga, and nadi-sodhana pranayama, during surya-namaskaram and during the practice of other yoga asanas described earlier. Ujjayi breathing will enhance the benefit, when combined with such practices.

Ujjayi breathing is similar to the action of smelling a flower, for example.

Sit comfortably and take a deep breath in, with both nostrils; notice the back of the mouth opening up during the slow and deep breathing. Then partially close the back of the throat with the *epiglottis*, similar to the swallowing action, and breathe out with force; notice the 'hissing' or 'snoring' sound the expiration makes.

Raising the tongue up to the palate and pulling the chin and abdomen inwards also help to close the throat partially.

Next try the Ujjayi breathing, during inhalation also, and the same 'hissing' sound can be heard.

Now do the Ujjayi breathing in both inhalation and exhalation, after closing the glottis and trachea with epiglottis, as described above, with same duration for both inhalation and exhalation.

Ujjayi Pranayama with Kumbakha:

After partially closing the glottis with epiglottis, Inhale through both nostrils; retain the breath with Jalandhara bandha or chin-lock, described previously under 'Kumbakha'; release the chin-lock; then exhale through the left nostril, by closing the right nostril with the right thumb.

What is Epiglottis?

Epiglottis is a flap-like cartilage, covered with mucus membrane, and is situated in the pharynx, behind the root of the tongue, above the *larynx*. In normal breathing, the epiglottis is up and relaxed, allowing the breathing to take place through the trachea. While swallowing food or liquid, the epiglottis is pushed down by elevation of the neighboring *hyoid bone*. This makes the epiglottis to close the *glottis and trachea*, preventing food or liquid to enter the trachea. It is the epiglottis that is pushed down to partially close the trachea in Ujjayi breathing.

CHAPTER 2

ON MEDITATION

❖ ❖ ❖

Beauty is truth, truth beauty–.
That is all ye know on earth
and all ye need to know.

— JOHN KEATS

MEDITATION, AS A DISCIPLINE IS one of the most important aspects of mind management, because it helps to clear the *cob-webs* of thoughts, the mind had woven in the past. Very often, people complain that when they sit for meditation other thoughts come and disturb them and their thoughts take them for a ride and for many, it is very difficult to bring the concentration back to meditation. The reason is, the mind itself consists of a *flow of thoughts*, like a chain; one thought is connected to the next thought and the next thought to the next and to the next and so on.

Dr. Rajah Sekaran

COBWEBS AND WEEDS OF THE MIND

In one fell-swoop, cut asunder
The cobwebs of the mind
Often in the midst of the weeds
Of mis-steps and past deeds
Tenacious with deep roots,
Using the sword of Chetana yoga
Of *Being Aware*, inherent in *Life Energy*
With the foundation pillars of
Kindness with good intentions,
Compassion and love,
Forgiveness and freedom,
And enjoy the *Peace;*
lest you fall off the cliff of freedom
Into the valley of arrogance and violence
Physical and mental,
And the perils of prejudice,
Practice, practice
Acts of kindness with good intentions,
Compassion and love;
For, the second and third attempts
Become harder and harder,
As the roots of the weeds
Become thicker and thicker.

Being aware of the flow of the thoughts, as they occur, will *empty out* the *cob-webs of thoughts* which have accumulated. It is best to ignore the flow of thoughts. The more and more the meditator is established in *Awareness*, similar to the *Chetana Yoga* described in the beginning chapters of this book, it will be less and less of a problem. With more practice in meditation, it will become easier to manage such thoughts.

The answer is to bring the focus back by being aware of the flow of thoughts from the heart center, the seat of kindess, compassion, and forgiveness and not through the mind, the seat of prejudice, anger, vanity, arrogance, and fear. The seat of being aware is the heart center, the subtle center of reality that gives origin to kindess, compassion, and forgiveness.

Pranayama and Meditation

It is best to do meditation after a brief session of Pranayama; however, one need not become proficient in Pranayama, before starting meditation. A few deep breaths (aka) *yogic breathing* described in the previous chapter on Pranayama, will help to bring the mind to a calm state before meditation.

A brief description is given on different meditation methods that can be helpful in keeping the focus on meditation:

1. MEDITATION BASED ON OBJECTS TO OBSERVE:

Objects to observe can be either a lamp or candle, so that the meditator can focus on the light, to develop concentration in the beginning. Here, there is subject-object relationship, until the person can go on to deeper meditation without the help of the objects.

2. MEDITATION BASED ON BREATHING:

In this method, sit comfortably and merely observe *normal breathing*, inhalation and exhalation. With practice, this observation will bring your mind to a calm status, where you can stay and continue meditation. As the observation or Awareness of breathing continues, the breathing slows, leading to deeper meditation, where Awareness alone remains.

3. MEDITATION BASED ON MANTRA WORD– MANTRA YOGA:

When a word or sentence–for example, OM (AUM)– is repeated mentally, the mind becomes calmer. When the same Mantra is repeated several times, there is 'silence' between the Mantra. The silence that lies between the Mantra word or words, becomes longer and longer, and eventually makes the mind *abide in the silence*. The word silence is used here for lack of a better word to denote the Changeless Truth which is the ultimate goal of meditation. In fact, all yoga systems are formulated to lead towards this ultimate goal.

4. MEDITATION BASED ON NADIS AND CHAKRAS–KUNDALINI YOGA:

Nadis are *subtle* nerve-like structures and they are distributed all over the body. Different ancient Yoga texts state different numbers for the total number of nadis to be over 300,000; however, it is generally considered to be a total of 72,000 nadis. Pranic or life-energy flows through them. When the flow is blocked, Life energy or Pranic energy cannot adequately flow through the nadis, giving rise to diseases and disturbance in the normal internal environment.

There are three main nadis along the spine; they are *Sushumna nadi, Ida nadi and Pingala nadi*. Sushumna Nadi is the most important and runs in the middle of the spinal column from the base of the spine to the top of the head. The Energy-Conciousness or the Pranic Energy that is concentrated in the lowest part of Sushumns nadi is called the *Kundalini Sakti* and is said to be coiled in three and a half circles like a snake. It is supposed to be lying dormant, until it rises up in the Sushumn nadi in the spine.

The Ida nadi runs to the left of the spinal column and the Pingala nadi runs to the right of the spinal column. They are also *subtle energy nadis*. Generally speaking, Ida nadi represents the para-sympathetic system and Pingala nadi represents the sympathetic system. They intertwine with each other at different centers from *Mooladhara Chakra* to the *Ajna Chakra* and merge with the Sushumna nadi in the *Ajna Chakra*, situated between the eyebrows.

Chakras are *subtle centers of Energy-Consciousness*. They are not anatomical structures but they are similar to the nerve-plexusus of the body distributed along the midline from the base of the spine to the top of the head. They are centers of Life energy or Pranic energy situated in the center of spine. There are mainly seven chakras and there are also other minor chakras. Each chakra has a lotus-like arrangement of petals; and the petals vary in number for each chakra. The number of petals represent the number of subtle nadis, arising from them. Each chakra, except for the *Sahasrara*, represents a *bija mantra (seed mantra)*. The lower five Chakras are associated with the five elements of space, air, fire, water and earth. *Mooladhara Chakra* represents the earth element.

A brief description of the seven main chakras are given here as follows:

Mooladhara Chakra is at the lower end of the Sushumna nadi, in the pelvic floor. In women it is said to be in the uterine cervix.

It is associated with the sacral plexus in the human body. It has four petals; mantra is *Lam;* the element is earth.

Swadhistana Chakra is about two inches behind the Mooladhara chakra and is associated with sacral and prostatic plexuses. It has six petals; mantra is *Vam*; the element is water.

Manipura Chakra is in the spinal column, at the level of the navel and is associated with solar plexus. It has ten petals; mantra is *Ram*; the element is fire.

Anahata Chakra is situated close to the heart and associated with cardiac plexus. It has twelve petals; mantra is *Yam*; the element is air.

Vishuddhi Chakra is in the middle of the throat. It is associated with cervical plexus in the neck area. It has sixteen petals; mantra is *Ham; the element is space.*

Ajna Chakra is behind the center of the eyebrows at the top of the spinal column and is a very important landmark in the practice of meditation. It is associated with mind and intelligence and is the ruler of all the lower centers. It has two petals; mantra is *Om*.

Sahasrara Chakra is the seventh and the highest of the major chakras. It is situated in the crown of the head, at the level of the anterior fontanelle. It is associated with the causal body or the *Most-Subtle* state.

Among the minor chakras, *Lalana Chakra*, situated in the back of the throat, opposite the epiglottis, is an important chakra. Purification of the Nadis and Chakras, with the practice of different asanas and Pranayama is an important preliminary step for meditation on Chakras as well as for general well-being of the human body.

It is said the microcosm is similar to and the prototype of macrocosm. In that sense, the human body, being the microcosm of the *cosmos*, the lower five chakras up to *Vishuddhi* represent the five elements of space, air, fire, water and earth in the *body*.

The *Ajna Chakra* represents the *subtle body* of the microcosm, the human body.

An example of the subtle body of the cosmos, in the present-day knowledge, would be *the Dark Matter and Dark Energy*, the origin and functioning of which are still unclear, and are being extensively investigated by *Physicists* and scholars of other disciplines.

The Sahasrara Chakra in this context, would be the *Most-Subtle Body* of the microcosm, the human body. The Most-Subtle Body is also called the *Causal body* by *Philosophers*.

The Absolute Changeless Truth, the ultimate goal of all the seekers of wisdom, in this author's opinion, is beyond the Least Subtle, the Subtle and the Most-Subtle of the *macrocosm as well as the microcosm*.

After gaining knowledge of the Chakras, the meditator is advised to raise Awareness from the lower to the higher chakras one by one, very gradually, to realize the Changeless Truth, ultimately.

This meditation will require guidance from a teacher who is familiar with the method.

CHETANA YOGA MEDITATION

Chetana yoga meditation is more or less a combination of the different types of meditation, described above. The major difference is, that the other types of meditation are related to the mind, with mind as the *end-point*. In Chetana yoga meditation, the meditator and the meditation are the *Awareness* itself; and the main focus is *Awareness*, in the beginning, during and at the end of meditation.

STEP ONE– CHETANA YOGA OF THE SURROUNDING:

Be Aware of the surrounding; observe the sounds coming from all around; and feel the breeze touching the skin. In other words, use your senses to go beyond the senses.

STEP TWO– CHETANA YOGA OF THE BODY:

Then close your eyes and close your ears with both hands, and observe the *OM* sound, also known as *Pranava*. *OM* is the origin of all sounds and words.

Then place the hands comfortably on your lap and observe the breathing; let the breathing be normal.

STEP THREE– CHETANA YOGA OF THE MIND:

Observe the mind; the flow of thoughts may be happening like a *chain* with one thought connected to the next thought and to the next thought etc,. Merely observe the chain of thoughts and watch them gradually subside in the process of observation. It will also help to remove the cobwebs of the mind, by being *Aware* of the disturbing thoughts and the flow of thoughts.

Use the *Mantra yoga*, described above, if needed at this time. Observe the *Silence*, between the words and gradually observe silence-within, without the words of the Mantra; and *be the silence*.

The silence between the words is the *Awareness, unlimited*. That is why, words fall short, when we attempt to describe Chetana- Awareness, unlimited.

THE ULTIMATE GOAL:
In any of the methods described, when the meditator becomes proficient, he is said to be in deep meditation, leading to the ultimate goal, which is for the mind to abide in Limitless Awareness.

When the mind ultimately abides in Limitless Awareness–being the *Origin of All Things in Nature*–it is called the Savikalpa Samadhi.

When the mind ultimately abides in Limitless Awareness–being the *Changeless Truth*–it is called the Nirvikalpa Samadhi.

There is no subject-object relationship; there is no meditator, object of meditation or process of meditation; there is no observer, observed and observation. The meditator is said to be established in "Steady wisdom".

EPILOGUE

❖ ❖ ❖

No man is an island entire of
itself; every man is a piece of the
continent, a part of the main;

— JOHN DONNE (1572-1631)

EVERY ONE OF US IS a part of the whole. The whole is the *All-pervasive mutually inherent Life Energy and Awareness*, described in this book as the origin of all things in Nature. It is beyond all functions of the mind including cognition, as seen in the human beings.

Even though the Big Bang happened 13 billion years ago, life as single-cell life, in the form of microorganisms started appearing about 3.7 to 4 billion years ago, after a cooling period of about 9 billion years after the Big Bang.

Life Energy inherent in Awareness existed even before the teleological development of a rudimentary brain with only the sense of smell, about 200 million to 300 million years ago in *Dinosaurs* and *Synapsids* to help them find food. *Synapsids*, resembling today's birds and reptiles lived about 320 million years ago.

A rudimentary brain and mind with sense of smell etc. probably developed sometime in that period, 200 million to 300 million years ago. The study of the fossils of Therapsida (Fossil TETRAPOD Order), from which mammals descended, 299 million to 200 million years ago, shows evidence of emergence of nasal turbinates, needed for the olfactory sense.

Synapsid Tetrapod, Brasilitherium, a predecessor of mammal, that lived 227 million years ago is also providing evidence of nasal turbiantes that facilitate the sense of smell.

So, it is clear that Life Energy existed long before the beginning of development of sense of smell and by inference, a rudimentary brain and its individualized mind needed to process the olfactory function.

Insects have evolved from 480 to 400 million years ago. Whereas, *homo-sapiens*, our ancestors came into existence only 300,000 years ago. So, Life-Energy with inherent Awareness-intelligence, as the basis of *existence*, existed for the insects, *dinosaurs, and synapsids*, resembling today's birds and reptiles, long before the beginning of human race and the beginning of mind.

Milela and Amilea

To understand our existence as a part of the whole, another acronym has been introduced earlier in the mind management of *Loneliness and Depression*, namely, *AMILEA*, meaning *All-pervasive Mutually Inherent Life Energy and Awareness*, to emphasize the all-pervasiveness of the whole.

MILELA was described earlier, in detail, as the very basis of our existence and the essence of our individual appearance and is the same *Life-Force* that is providing our individual body and mind. The Milela theory of inherence of Life Energy and Awareness is based on the *Epiphany* that the Awareness is *beyond mind* and is not a part of the mind, as it has been generally believed. The proof of the same has been dealt with in the earlier chapters of *epiphany and Milela Theory*. Mind still exists as a part of the body, based on the *all-pervasive Awareness*.

MILELA and *AMILEA* carry the *same meaning and is our true nature;* and forms the foundation for the practice of *CHETANA YOGA* of observing the individual mind, as an antidote for our mental frustrations and ambiguities, such as anger and anxiety and fear, and also in the prevention of *isolation, loneliness and depression*, as explained earlier, in the 'Mind management.'

Chetana Yoga–for Mind Management and as A Remedy for Isolation

Chetana Yoga, in short, is the technique of developing *a habit of being aware of our thoughts and actions as they occur.* The practice of Chetana Yoga, described earlier, in detail, in the chapter under the same name, creates a permanent layer between the thoughts and action, before the action is committed, enabling *equanimity* and balance to the mind. Equanimity or *samathvam* in Sanskrit is Yoga, so says Bhagavad Gita, as *"Samathvam yogam uchyate"* in one of the verses.

The practice of Chetana Yoga, based on the theory of AMILEA or MILELA–emphasizing the all-pervasive Awareness as our true nature–leads to a sense of belonging. Moreover, ignorance of this theory of all-pervasiveness of our inner-self leads one to a lack of equanimity. The lack of equanimity and a lack of sense of belonging leads one to *isolation, with a subjective feeling of isolation from others* and *loneliness* and eventually to *depression.*

So, the practice of *Chetana Yoga* is again emphasized as a necessity with the understanding that every one of us is a part of the whole, the *all-pervasive Awareness space*; and there is no justification in prejudice of various kinds, isolation and loneliness.

Amilea/Milela and Quantum Physics

All-pervasiveness of *Amilea* includes the Quantum world of subatomic particles which are basically *kinetic energy (Life-Energy)* in momentum in the *Awareness-space*. The potential Energy of Awareness space (the Most-Subtle) expands as Gravity space and Electro-Magnetism as the duality which co-exist continuously both in the subtle space and the least-subtle (gross) space. The least-subtle (gross) space includes our world. All of us are nothing but different combinations of the subatomic particles of Electro-Magnetism, dancing away in the all-pervasive Awareness-space, appearing as the body and mind. This is true of both animate and inanimate, appearing as the different forms.

Carl Sagan was right, when he said,

"Science is not only compatible with spirituality; it is a profound source of spirituality"

Science may be meaningless without spirituality; nevertheless, science provides the meaning and a path for the understanding of spirituality which is often esoteric.

THE PANDEMIC AND THE RISING OF ICEBERG:
it was described earlier in this book that the increasing incidence of gun violence, depression etc., is the 'Tip of the Iceberg'. Now in view of the recent global developments with the Novel Corona Virus, it is reasonable to assume the Iceberg itself is rising up, in the form of the Covid-19 Pandemic that shocked the world from December 2019.

However, there is increasing possibility the Corona SARS-2 virus causing the COVID-19 or a variation of the same may have been around long before the recent Pandemic. For example, a six-month old baby admitted for Kawasaki disease tested positive for COVID-19; and was reported in April 2020 from Lucille Packard Children's Hospital, Stanford, California. Since then, in early May 2020, fifteen more similar cases were reported by New York City Health Department and 64 cases statewide in New York were also reported with the diagnosis of "Pediatric Multi-system Inflammatory Syndrome-temporarily associated with Covid-19".

Europe also had reported a similar presentation of patients with the Syndrome, with Covid-19, among children. This multi-system inflammatory disease affects the heart also with vasculitis of the coronary arteries causing heart failure.

Whereas Kawasaki disease was described over 50 years ago, in 1967 in Japan. Hawaii Children's

Epilogue

Hospital reported the first case of Kawasaki disease in USA, with about 20 cases between 1974 and 1975. The cause was unknown.

It is now clear that Covid-19 is a multi-organ Inflammatory disease affecting heart, kidneys, lungs and brain. When it affects the brain, the patients have confusion and delirium. The main pathological finding in the lungs is again inflammation with swelling, profuse mucus production and abnormal immune response with 'Cytokine Storm'/ The treatment of the lungs, likewise differs from the traditional treatment of Acute Respiratory Distress Syndrome (ARDS). So, it was found by the Pulmonologists, that the Ventilator use with high pressure had caused more deaths; and it had to be tailored down to low pressure. Even the C-PAP MACHINE is enough to provide Oxygen supply to the drowning lungs.

For 300 years, various Flu and Swine-Flu epidemics have devastated the world population, including SARS in 2003, MERS in 2012 and now Corona SARS-2 in 2019 with the possibility of future pandemics. As soon as the humans control one epidemic, another one sprouts up sooner than the previous, like the villains of Mythology and Science fiction.

Time will tell as we unravel the mystery of the Novel Corona virus and the 'Rising Iceberg metaphor'- connection.

The New Normal

it is heart-wrenching to see hundreds of thousands of people dying world-wide and many, many more, suffering from COVID-19, with insurmountable extent of social problems and overwhelming economic devastation. However, it seems like this Covid-19 is marching on, as the *'compensatory feed-back mechanism'* of the world, correcting man-made problems of pollution of Land and Sea, and increasing incidence of gun violence, depression etc., as noted earlier. It is similar to the *'Internal milieau'* of the human body, maintaining the balance of the various systems, through the *Hypothalamico-Pituitary axis'*.

Inflammation seems to be the main pathology affecting one or more organs. So, *anti-inflammatory* drugs are being tried as one of the tools of treatment. At the same time, diet with anti-inflammatory ingredients should be considered on a daily basis, as a preventive measure, for the general public with or without COVID-19. Among the diet choices, spices such as *Turmeric, Cumin etc.*, are known to contain a very significant amount of anti-inflammatory and anti-oxidant properties; and should be made an integral part of the USDA-MY PLATE (known previously as Food-Pyramid) recommendation.

The 'New-Normal' appears to be leading us to our realization that 'We are all in this (World) together'. It is also gratifying to see that there is more *kindness and caring* and less arrogance and violence in the society at large; and overall better inter-relationships.

Epilogue

LOVE AND ATTACHMENT

Mind is from Mars; Love is from Venus.
Attachment arises from Mind, whereas,
Heart Center is the seat of Love and Kindness,
Compassion and Forgiveness.
Attachment is not Love,
Attachment is possessiveness, whereas,
Children and other loved-ones not our possessions.
So 'Accept' them as they are, with Love;
Guidance, of course, everyone needs.
Possessiveness leads to violence
As mental, physical and /or verbal abuse.
So 'let go' attachment and possessiveness;
'Let go' attachment to past events;
'Accept' the past events and future;
Have dreams, ambitions,
Planning and expectations and
Act with full effort, dexterity and commitment;
But 'accept' the outcome; for it is not in our control.
Attachment is from mind,
Love is from Heart, Beyond Mind.

ACKNOWLEDGEMENT

SHANTHI SEKARAN PHD HOLDS A doctorate degree in Creative Writing and teaches Fiction Writing at the college-level. I am grateful to Shanthi for teaching me the fundamentals of writing as well as editing some of the chapters in this book in the early period of my writing.

Spencer Dutton PhD, Shanthi's husband, is a Physicist at the Lawrence Berkeley National Lab and has been my inspiration to begin with, by providing me Physics articles that suggested, other disciplines of Philosophy and Mysticism may throw a light on some of the unanswered questions in Physics. I am thankful to Spencer.

I am grateful to Sunaina Sekaran and Nikhil Sekaran for typing and editing along with me in the recent period of finishing the manuscript.

I am thankful to Alan Freeland. Alan Freeland is an attorney. Alan and I have had many discussions at the Berkeley Tennis Club over coffee on several topics including Mind and Mysticism. Those discussions inspired me to start writing this book.

I am also thankful to my teachers of various branches of Philosophy of India and Kriya Yoga, most notably, Swami Dayananda Saraswati, Swami Shraddhananda and Vedathiri Maharishi.

Elite Editing is an independant agency, offering all facets of editing; and did a thorough evaluation of the title and the subject material of the book to come up with the cover design and in the completion of the final editing of the manuscript. I am thankful and I also appreciate the patience and guidance by Jennifer Chandler, the publishing consultant. I thank Lydia Bowman, the Project Manager who was immensely helpful in bringing the 'Revised second edition' to fruition.

Last, but not the least, I am especially thankful to my wife, Surabhi Sekaran M.D. who gave me moral support to continue to write in the past few years and to put together various topics to bring out this book.

GLOSSARY

SANSKRIT WORDS:

Abhaya Mudra	Denotes the message, 'fear not'
Advaita	Non-Duality
Ajna Chakra	The sixth of seven *Chakras*, it is an important landmark in the practice of meditation, associated with mind and intelligence
Ākarshana	Attracting; Gravity force
Ākarshika	Both attractive force and magnetic force
Akasha	Space
Ākrishti	Gravity force
Anahata Chakra	The fourth of the seven *Chakras*, is associated with the cardiac plexus
Ananda	Bliss, or happiness
Anatha	A *Pali* language word for no individual soul, egolessness
Anāthma	No individual soul, egolessness
Anatthama	No individual soul, egolessness
Antakaranam	Inner instrument
Bastrika	Bellow-like breathing
Bija Mantra	Seed-mantra
Chaitanya	Awareness or Consciousness
Chakras	Subtle centers of Energy Consciousness, similar to nerve plexuses

Chetana	Awareness or Consciousness
Chit	Awareness or Consciousness
Chita	Awareness or Consciousness
Chith	Awareness or Consciousness
Durga	One of many names given to the Mother Goddess
Grantham	Belongs to both Tamil and Sanskrit; refers to the Sanskrit text written in Tamil script
Ida nadi	Subtle *nadi* running through the left of the *Sushumna nadi*
Jalandara bandha	Chin-lock
Kānta Shakti	Magnetic Energy
Kanti	Brilliance
Kapalabhati	A type of *Pranayama*, done at the level of the back of the nose
Karma-Palan	Fruits of action
Kumbhaka	Retention of the breath
Kundalini Shakti	Subtle Energy, lying dormant in the base of the *Sushumna Nadi*
Lalana Chakra	A minor *Chakra*, situated in the back of the throat
MahaKali	One of many names given to the Mother Goddess
Mahabharata	A great Hindu Epic
Manipuraka Chakra	The third of the seven Chakras, is associated with the solar plexus

Glossary

Mooladhara	The first of the seven *Chakras*, *Chakra* situated in the midline of the body, is associated with the sacral plexus in the human body
Nadi	Subtle nerve-current resembling the nerve of the nervous system
Nataraja	Dancing *Siva*
Nirvikalpa -Samadhi	State of merging with the Ultimate Truth
Pali	A dialect of Sanskrit, used by common people in the kingdom of Kapilavastu at the time of Gautama Buddha
Pandavas	Sons of King Pandu, belonged to the Kuru Dynasty in *MahaBharata*
ParaShakti	One of many names given to the Mother Goddess
Parvati	One of many names given to the Mother Goddess
Pingala nadi	Subtle *nadi* running through the right of the *Sushumna nadi*
Prana	Life-force; breathing
Pranayama	Control of breathing in order to control the mind
Puraka	Breathing in or inhalation

Rechaka	Breathing out or exhalation
Sahasrara Chakra	The seventh of the seven *Chakras*, is associated with the Most-Subtle State
Samadhi	State of merging with the ultimate Reality or Truth
Samatvam	Equanimity
Sanyāsi	A Hindu renunciate
Sat	Truth or Existence
Sat-Chit-Ananda	Awareness or Consciousness
Savikalpa-Samadhi	State of merging with the ultimate Reality
Seetali	A *Pranayama* method to cool the body; also known to lower blood pressure
Siddhi	Success
Siva	Represents the manifestation of Shakti, the potential Energy of the Supreme Being, also known as *Sivam*
Sivam	The Formless Supreme Being
Sukshmam	Subtlety
Sushumna nadi	Subtle *nadi* running through the middle of the spinal column
Swadistana Chakra	The second of the seven *Chakras*, is associated with the sacral and prostatic plexuses of the human body

Glossary

Swarupa	Inner nature
Ujjayi breathing	A type of *Pranayama*
Vipareeta Svasa	A type of *Pranayama* done at the level of the tip of the nose
Vishuddhi Chakra	The fifth of the seven *Chakras*, is associated with the cervical plexus
Vāsanās	Based on actions, as well as unfulfilled strong desires with attachment and cravings of the mind from the previous birth
Yama	Control
Yoga	To join, to yoke; from the root yuj, commonly used to describe different techniques used towards the ultimate goal of joining the individual self with the supreme self
Yogi	A Practitioner of Yoga

TAMIL WORDS:

Bootham	All things; generally refers to space, air, fire, water and earth
Kesari	A sweet made from cream of wheat
Māmatha yanai	Wild elephant
Marainthathu	Hidden; not seen
Maraithal	Obscure
Maram	Wood
MunVinai	Unfulfilled fruits of actions and desires with strong attachments from previous birth
Param	The highest reality
Parmuthal	In the world
Pongal (sweet)	A sweet made from rice
ThiruMoolar	A mystic and poet who lived 3000 years ago in South India
Thiruachi	The circular structure seen around the deity representing the Consciousness or Awareness

LATIN WORDS:

Cogito	I think
Ergo	Therefore
Sum	I am (I exist)

OTHER WORDS:

Ego	Provides the personality of the person based on his or her self impression of one's past achievements, thoughts, memories, future plans and ambitions
Milieu	Environment
Plato	An ancient Greek philosopher, disciple of Socrates

ACRONYMS:

AMILEA	All-Pervasive Mutually Inherent Life Energy and Awareness
MILELA	Mutually Inherent Life Energy and Limitless Awareness

SELECTED BIBLIOGRAPHY

Aizenman, Nurith. "Gun Violence: Comparing The U.S. With Other Countries." *National Public Radio*, November 6, 2017.

ArunaiVadivel Mudaliar. *ThiruMantiram by Thirumoolar.* (In Tamil) Mayuram, Dharmapuram Adheenam, 1995.

Azhagaradigal. *Kandar Anubhuti of Saint Arunagirinathar.* Chennai, Thondaman Chakravarti Publishers, 1995.

Bonnie Berkowitz, Denise Lu, and Chris Alcantara. "The Terrible Numbers that Grow with Each Mass Shooting." *The Washington Post*, Updated June 29, 2018.

Brandon, Anja Malawi, Sahar H. El Abbadi, Uwakmfon A. Ibekwe, Yeo-Myoung Cho, Wei-Min Wu, Craig S. Criddle. "Fate of Hexabromocyclododecane (HBCD), A Common Flame Retardant, In Polystyrene-Degrading Mealworms: Elevated HBCD Levels in Egested Polymer but No Bioaccumulation." *Environmental Science & Technology* 2019.

Capra, Fritjof. *The Web of Life.* New York, Anchor Books, October 1997.

Carney, D. R., A. J. C. Cuddy, and A. J. Yap. "Power Posing: Brief Nonverbal Displays Affect Neuroendocrine Levels and Risk Tolerance." *Psychological Science* 21.10 (2010): 1363-368. Web.

De Prado Bert, P., Mercader, E.M.H., Pujol, J. et al. "The Effects of Air Pollution on the Brain: a Review of Studies Interfacing Environmental Epidemiology and Neuroimaging." July 14, 2018.

Debra M. Stone, Thomas R. Simon, Katherine A. Fowler, Scott R. Kegler, Keming Yuan, Kristin M. Holland, Asha Z. Ivey-Stephenson, Alex E. Crosby. "Vital Signs: Trends in Suicide Rates—United States, 1999-2016 and Circumstances Contributing to Suicide—27 States, 2015." *Centers for Disease Control and Prevention*, June 8, 2018.

"Alone in the Crowd" *The Economist*, September 1, 2008, pages 49-51.

"The Oldest Homo Sapiens Yet." *The Economist*, June 10, 2017.

Geuze, et al. "Reduced GABA Benzodiazepine Receptor Binding in Veterans with Post-Traumatic Stress Disorder." Molecular Psychiatry, 2008, pages 74-83.

Gould, Skye and Mosher, Dave. "How likely is gun violence to kill the average American? The odds may surprise you." *Business Insider*, February 15, 2018.

Karthikeyan, N.V. *Kandar Anubhuti of Saint Arunagirinathar.* Shivanandanagar, India, The Divine Life Society, 1990.

Lamm, Ben. "Algae might be a secret weapon to combating climate change." QUARTZ October 1, 2019.

Master Subramuniya. *Raja Yoga.* San Francisco, Comstock House, 1973.

Morris, Sam and Guardian Interactive Team. "Mass shootings in the US: there have been 1,624 in 1,870 days." The Guardian, February 15, 2018.

Press Release: Suicide Rates Rising Across the U.S. *Centers for Disease Control and Prevention*, June 7, 2018.

Ramanatha Pillai. *ThiruMantiram by Thirumoolar.* (In Tamil) Chennai, The South India Saiva Siddhanta Works Publishing Society, 1980.

Saravanananda. *Arut Perum Jothi Agaval.* (In Tamil) Chennai, Ramalingar Panimandram Publishers, 1974.

Schmidt, Michael. "F.B.I. Confirms a Sharp Rise in Mass Shootings Since 2000." *The New York Times*, September 24, 2014.

St. James, Elaine. *Inner Simplicity.* New York, Hyperion, 1995.

Streeter, Chris C., MD. "Yoga Asana Sessions Increase Brain GABA Levels." *The Journal of Alternative and Complementary Medicine* 13.4 (2007): 419-26. Print.

Sugarman, Joe. "The Rise of Teen Depression." *Johns Hopkins Review*, Fall/Winter 2017, Volume 4, Issue 2.

Swami Chidbhavananda. *The Bhagavad Gita.* Tamil Nadu, Sri Ramakrishna Tapovanam, Tirupparaitturai, 1986.

Selected Bibliography

Swami Chinmayanda. *Self-Unfoldment.* Piercy, CA, Chinmaya Publications, 1992.

Swami Chinmayananda. *The Holy Geeta.* Bombay, Sri Ram Batra, Central Chinmaya Mission Trust, 1960.

Swami Dayananda Saraswati. *Bhagavad Gita.* Saylorsburg, PA, Arsha Vidya Gurukulam Institute of Vedanta & Sanskrit.

Swami Gambhirananda. *Eight Upanisads.* Mayavati, Pithoragarh, Himalayas, Advaita Ashrama, August 1978.

Swami Nikhilananda. *The Gospel of Sri Ramakrishna.* New York, Ramakrishna -Vivekananda Center, 1980.

Swami Nikhilananda. *The Mandukyopanisad.* Mysore, The President Sri Ramakrishna Ashrama, 1974.

Swami Nihsreyasananda. *Man and His Mind.* Mylapore, Madras, Sri Ramakrishna Math, April 25, 1993.

Swami Saravanananda. *Aruperunjothi Agaval.* (In English) Madras, Ramalinga Mission.

Terry, Robert M.D. et al. "Physical basis of cognitive alterations in Alzheimer's Disease: Synapse loss is major correlate of cognitive impairment." *Annals of Neurology*, Vol. 30, Issue 4, 1991.

Thulasiram, T.R. *Arut Perum Jothi and Deathless Body–a study of Swami Ramalingam*. Madras, University of Madras, 1980.

Varadarajan, G. *ThiruMantiram by Thirumoolar.* (In Tamil) Chennai, Palaniappa Brothers, 1985.

Wilson, Chris. "35 Years of Mass Shootings in the U.S. in One Chart." *Time Magazine*, Updated November 5, 2017.

Zukav, Gary. *The Dancing Wu Li Masters.* New York, Bantam Books, August 1980.

www.ingramcontent.com/pod-product-compliance
Lightning Source LLC
Chambersburg PA
CBHW032029290426
44110CB00012B/732